TOPICS IN ALMOST AUTOMORPHY

TOPICS IN ALMOST AUTOMORPHY

Gaston M. N'Guérékata

Morgan State University
Baltimore, Maryland

 Springer

Library of Congress Cataloging-in-Publication Data

N'Guérékata, Gaston M., 1953–
   Topics in almost automorphy/Gaston M. N'Guerekata.
      p.   cm.
   Includes bibliographical references and index.
   ISBN 0-387-22846-2
   1. Automorphic functions.   I. Title

QA353.A9N52 2004
515'.9—dc22
                                                2004059527

©2005 Springer Science+Business Media, Inc.
New York, Boston, Dordrecht, London, Moscow

ISBN 0-387-22846-2 (Hardbound)   Printed on acid-free paper.

Printed in the United States of America.   (BS/DH)

9 8 7 6 5 4 3 2 1   SPIN 11305217

springeronline.com

In memory of Thérèse N'Guérékata, my mother

Preface

Since the publication of our first book [80], there has been a real resurgence of interest in the study of almost automorphic functions and their applications ([16, 17, 28, 29, 30, 31, 32, 40, 41, 42, 46, 51, 58, 74, 75, 77, 78, 79]). New methods (method of invariant subspaces, uniform spectrum), and new concepts (almost periodicity and almost automorphy in fuzzy settings) have been introduced in the literature. The range of applications include at present linear and nonlinear evolution equations, integro-differential and functional-differential equations, dynamical systems, etc...It has become imperative to take a bearing of the main steps of the theory.

That is the main purpose of this monograph. It is intended to inform the reader and pave the road to more research in the field. It is not a self contained book. In fact, [80] remains the basic reference and fundamental source of information on these topics.

Chapter 1 is an introductory one. However, it contains also some recent contributions to the theory of almost automorphic functions in abstract spaces.

Chapter 2 is devoted to the existence of almost automorphic solutions to some linear and nonlinear evolution equations. It contains many new results.

Chapter 3 introduces to almost periodicity in fuzzy settings with applications to differential equations in fuzzy settings. It is based on a work by B. Bede and S. G. Gal [40].

Finally in Chapter 4 the classical theory of almost automorphic vector-valued functions is extended to fuzzy settings. This chapter begins with the presentation of several "new" spaces in which the theory holds, called fuzzy-number type spaces. These spaces are more general than the Banach and Fréchet spaces, since they are not linear structures although they present nice metric properties. Their importance consists in the fact that they are very appropriate for situations where imprecision which appears in the modelization of real world problems by differential equations is due to uncertainty or vagueness (and not randomness). Applications to some fuzzy differential equations are also given. It is based on S. G. Gal and G. M. N'Guérékata's recent work [41].

At the end of each chapter, we recall some relevant bibliographical remarks and raise some open problems and/or potential research subjects for graduate students and begining researchers in the area. It is our hope that this monograhp be used to stimulate some seminars and graduate courses in Analysis, Dynamical Systems, Fuzzy Mathematics and other branches of Mathematics.

Ackowledgements. I like to express my deepest gratitude to my colleagues and friends Professors D. Bugajewski and S. G. Gal who gave the entire manuscript a careful proofreading. Their comments and valuable suggestions have been helpful while I have been selecting the topics of this monograph. I also appreciate collaborating with Professors Nguyen Van Minh, Jerome A. Goldstein and James Liu over the past 2 years.

I would like to express my appreciation for the editorial assistance I received from Kluwer, especially from Ana Bozicevic. Many thanks to Morgan State University officials for granting me the necessary financial support during the preparation of the manuscript.

Finally, this book would hardly have been possible without the emotional support and encouragement of my wife Béatrice.

Baltimore, MD- USA   *Gaston M. N'Guérékata*

                                                      May 2004

Contents

1

Introduction and Preliminaries

This chapter has an introductory character to this monograph. We wish to recall briefly some concepts, results, methods and notations that will be used in the sequel. We will indicate in general some references where the reader can find more informations if necessary. Although for almost automorphy, our book [80] remains the main source of information, we give detailed proofs to some new results.

1.1 Measurable Functions

In this section we will recall some facts about measurable vector-valued functions and their integrals. We consider $(X, \|.\|)$ a Banach space and I an open interval in $\mathbb{R}$. We denote by $C_c(I; X)$ the Banach space of continuous functions $f : I \to X$ with compact support in I.

Definition 1.1. *A function $f : I \to X$ is said to be measurable if there exists a set $S \subset I$ of measure 0 and a sequence $(f_n) \subset C_c(I; X)$ such that $f_n(t) \to f(t)$ as $n \to \infty$, for all $t \in I \backslash S$.*

We observe that if $f : I \to X$ is measurable, then $\|f\| : I \to \mathbb{R}$ is measurable too.

Theorem 1.2. *Let $f_n : I \to X$, $n = 1, 2...$ be a sequence of measurable functions and suppose that $f : I \to X$ and $f_n(t) \to f(t)$ as $n \to \infty$, for almost all $t \in I$. Then f is measurable.*

Proof. We have $f_n \to f$ on $I \backslash S$, where S is a set of measure 0. Let $(f_{n,k})_{k \in \mathbb{N}}$ be a sequence of functions in $C_c(I; X)$ such that $f_{n,k} \to f_n$ almost everywhere on I as $k \to \infty$. By Egorov's Theorem (see [90, p.16]) applied to the sequence of functions $\|f_{n,k} - f_n\|$, there exists a set $S_n \subset I$ of measure less than $\frac{1}{2^n}$ such that $f_{n,k} \to f_n$ uniformly on $I \backslash S_n$, as $k \to \infty$.

Now let $k(n)$ be such that $\|f_{n,k(n)} - f_n\| < \frac{1}{n}$ on $I \backslash S_n$ and $F_n = f_{n,k(n)}$. Also let $B = S \bigcup (\bigcap_{m \geq 1} \bigcup_{n > m} S_n)$. Then it is clear that B is a subset of I of measure 0. Take $t \in I \backslash B$. So we get $f_n(t) \to f(t)$, as $n \to \infty$. On the other hand if n is large enough, $t \in I \backslash S_n$. It follows that $\|F_n - f_n\| < \frac{1}{n}$. Which means $F_n(t) \to f(t)$, as $n \to \infty$, and consequently, f is measurable. $\square$

Remark 1.3. It is easy to observe that if $\phi : I \to \mathbb{R}$ and $f : I \to X$ are measurable, the $\phi f : I \to X$ is measurable too.

Theorem 1.4. *(Pettis' Theorem) A function $f : I \to X$ is measurable if and only if the following two conditions are satisfied:*

(a) f is weakly measurable (i.e. for every $x^* \in X^*$, the dual space of X, the function $(x^* f)(t) : I \to \mathbb{R}$ is measurable)

(b) There exists a set $S \subset I$ of measure 0 such that $f(I \backslash S)$ is separable.

Proof. See [90, p.131]. □

We also have the following

Theorem 1.5. *If $f : I \to X$ is weakly continuous, then it is measurable.*

Definition 1.6. *A measurable function $f : I \to X$ is said to be integrable on I if there exists a sequence of functions $f_n \in C_c(I; X)$, $n = 1, 2, \ldots$ such that*

$$\int_I \|f_n(t) - f(t)\| dt \to 0, \quad as \ n \to \infty.$$

Remark 1.7. If $f : I \to X$ is integrable, it can be shown that there exists a vector $x \in X$, such that if $f_n \in C_c(I; X)$, $n = 1, 2, \ldots$ and $\int_I \|f_n(t) - f(t)\| dt \to 0$ as $n \to \infty$, then $\int_I f_n \to x$ as $n \to \infty$. Such x is called the integral of f on I and denoted $x := \int_I f$.

Moreover if $I = (a, b)$, then we denote $x := \int_a^b f$.

Theorem 1.8. *(Bochner's Theorem). Assume $f : I \to X$ is measurable. Then f is integrable if and only if $\|f\|$ is integrable. Moreover we have*

$$\left\| \int_I f \right\| \le \int_I \|f\|.$$

Proof. Let $f : I \to X$ be integrable. Then by the definition, there exist $f_n \in C_c(I; X)$, $n = 1, 2, \ldots$ such that $\int_I \|f_(t) - f(t)\| dt \to 0$ as $n \to \infty$.

We have $\|f\| \le \|f_n\| + \|f_n - f\|$, for each n, so $\|f\|$ is integrable.

Conversely assume now $\|f\|$ is integrable. Let $F_n \in C_c(I; \mathbb{R})$, $n = 1, 2, \ldots$ be a sequence of functions such that $\int_I |F_n(t) - \|f(t)\|| dt \to$

0 as $n \to \infty$ and $|F_n| \leq F$ almost everywhere for some $F : I \to \mathbb{R}$, with $\int_I \|F(t)| dt < \infty$.

Since f is measurable, there exist $f_n \in C_c(I; X)$, $n = 1, 2, ...$ such that $f_n \to f$ almost everywhere.

We now let

$$u_n = \frac{|F_n|}{\|f_n\| + \frac{1}{n}} f_n, \quad n = 1, 2, ...,$$

then it is obvious that $\|u_n\| \leq F$ for each $n = 1, 2...$, and $u_n \to f$ almost everywhere on I. Therefore $\int_I \|u_n - f\| dt \to 0$ as $n \to \infty$ and so f is integrable.

Using Lebesgue-Fatou's Lemma (see [90]), we get

$$\| \int_I f \| \leq \lim_{n \to \infty} \| \int_I u_n \|$$

$$\leq \lim_{n \to \infty} \int_I \|u_n\|$$

$$\leq \int_I \|f\|.$$

The proof is complete.   □

Theorem 1.9. *(Lebesgue's Dominated Convergence Theorem). Let $f_n : I \to X$, $n = 1, 2, ...$ be a sequence of integrable functions and $g : I \to \mathbb{R}$ be an integrable function. Let also $f : I \to X$ and assume that:*

(i) for all $n = 1, 2, ...$ $\|f_n\| \leq g$, almost everywhere on I, and
(ii) $f_n(t) \to f(t)$, as $n \to \infty$ for all $t \in I$.

Then f is integrable on I and

$$\int_I f = \lim_{n \to \infty} \int_I f_n.$$

Definition 1.10. *Let* $1 \leq p \leq \infty$. *We will denote by* $L^p(I; X)$ *the space of all classes of equivalence (with respect to the equality a.e on* I*) of measurable functions* $f : I \to X$ *such that* $\|f\| \in L^p(I)$.

If we define a norm on $L^p(I; X)$ *by*

$$\|f\|_p = (\int_I \|f(t)\|^p dt)^{\frac{1}{p}}, \quad if \ 1 \leq p < \infty$$

and

$$\|f\|_p =: \|f\|_\infty = ess \sup_I \|f(t)\|, \quad if \ p = \infty,$$

then $L^p(I; X)$ *is a Banach space.*

We shall denote by $L^p_{loc}(I; X)$ the space of all (equivalence classes of) measurable functions $f : I \to X$ such that the restriction of f to every bounded subinterval of I is in $L^p(I; X)$.

1.2 Sobolev Spaces

Let $\Omega \subset \mathbb{R}^n$ be an open bounded subset.

Definition 1.11. *A function* $g \in L^1_{loc}(\Omega)$ *is said to be the weak derivative of a function* $f \in L^1_{loc}(\Omega)$ *(or a derivative in the sense of distributions of order* α*), if*

$$\int_\Omega g \cdot \phi \ dx = (-1)^{|\alpha|} \int_\Omega f \cdot D^\alpha \phi \ dx, \quad for \ all \ \ \phi \in C_0^\infty(\Omega).$$

In this case, we write $g = D^\alpha f$.

Recall that $D^\alpha f$ denotes the α-derivative defined by:

$$D^\alpha f := \frac{\partial^{|\alpha|} f}{\partial x_1^{\alpha_1} \partial x_2^{\alpha_2} ... \partial x_n^{\alpha_n}},$$

where $\alpha = (\alpha_1, \alpha_2, ..., \alpha_n)$, α_i $(1 \leq i \leq n)$ is a nonnegative integer, and $|\alpha| = \alpha_1 + \alpha_2 + ... + \alpha_n$.

Definition 1.12. *Let k be a non-negative integer and let $1 \leq p \leq \infty$. We define the Sobolev space $W^{k,p}(\Omega)$ by*

$$W^{k,p}(\Omega) := \{f \in L^p(\Omega) : \quad D^\alpha f \in L^p(\Omega) \quad \text{for all } |\alpha| \leq k\}.$$

In $W^{k,p}(\Omega)$, we define a norm by

$$(N)_{k,p} \quad \|f\|_{k,p}^p := \int_\Omega \sum_{|\alpha| \leq k} |D^\alpha f(x)|^p dx, \quad p < \infty,$$

$$(N)_{k,\infty} \qquad\qquad \|f\|_{k,\infty} := \max_{|\alpha| \leq k} \|D^\alpha f\|_\infty,$$

and for $p = 2$, we define an inner product

$$(N)_{k,2} \qquad\qquad \langle f, g \rangle := \sum_{|\alpha| \leq k} \int_\Omega \overline{D^\alpha f} \cdot D^\alpha g \, dx.$$

We have the following

Theorem 1.13. *The space $W^{k,p}(\Omega)$ is a Banach space. If $p < \infty$, it is separable.*

Also we have

Definition 1.14. *By $W_0^{k,p}(\Omega)$ we denote the closure of $C_0^\infty(\Omega)$ in $W^{k,p}(\Omega)$. ($C_0^\infty(\Omega)$ denotes the space of functions of class C^∞ with compact support in Ω.*

Definition 1.15. *We define $H^k(\Omega) := W^{k,2}(\Omega)$ and $H_0^k(\Omega) := W_0^{k,p}(\Omega)$.*

Theorem 1.16. *$H^k(\Omega)$ and $H_0^k(\Omega)$ are Hilbert spaces when endowed with the inner product $(N)_{k,2}$.*

Theorem 1.17. *$C^\infty(\Omega) \cap H^k(\Omega)$ is dense in $H^k(\Omega)$, where $C^\infty(\Omega)$ is the space of functions defined on Ω of class C^∞.*

1.3 Semigroups of Linear Operators

Definition 1.18. *Let $A : X \to X$ be a linear operator with domain $D(A) \subset X$, in a Banach space $(X, \|\cdot\|)$. The family $T = (T(t))_{t \in \mathbb{R}^+}$ of bounded linear operators on X is said to be a C_0-semigroup if*

i) For all $x \in X$, the mapping $T(t)x : \mathbb{R}^+ \to X$ is continuous;

ii) $T(t + s) = T(t)T(s)$ for all $t, s \in \mathbb{R}^+$ (semigroup property);

iii) $T(0) = I$, the identity operator.

The operator A is called the infinitesimal generator (or generator in short) of the C_0-semigroup T if

$$Ax = \lim_{t \to 0^+} \frac{T(t)x - x}{t}$$

where

$$D(A) = \left\{ x \in X / \lim_{t \to 0^+} \frac{T(t)x - x}{t} \ exists \right\}.$$

It is observed that A commutes with $T(t)$ on $D(A)$. We define a C_0-group in a similar way, replacing $\mathbb{R}^+$ by $\mathbb{R}$.

For a bounded linear operator A, we have

$$T(t) := e^{tA} = \sum_{n=0}^{\infty} \frac{t^n A^n}{n!}.$$

We also have the exponential growth:

Proposition 1.19. *(see [90] page 232). If $T = (T(t))_{t \in \mathbb{R}^+}$ is a C_0-semigroup then there exist $K > 0$ and $\omega < \infty$ such that*

$$\|T(t)\| \leq Kc^{\omega t}, \ for \ all \ t \in \mathbb{R}^+.$$

If $\omega < \infty$, we say that T is exponentially stable.

Proposition 1.20. *If* $T = (T(t))_{t \in \mathbb{R}^+}$ *is a* C_0-*semigroup, then:*

a) the function $t \to \|T(t))\|$, $\mathbb{R}^+ \to \mathbb{R}^+$ *is measurable and bounded on any compact interval of* $\mathbb{R}^+$.

b) the domain $D(A)$ *of its generator is dense in* X.

c) the generator A *is a closed linear operator.*

1.4 Fractional Powers of Operators

Let $(X, \|.\|)$ be a (complex) Banach space and let $C : D(C) \subset X \mapsto X$ be a densely defined closed unbounded linear operator acting in X. Assume that $-C$ is the infinitesimal generator of an analytic semigroup $(R(t))$ and that $0 \in \rho(C)$, where $\rho(C)$ is the resolvent of the operator C. Then one can define, for $0 < \alpha \leq 1$, the fractional powers of C^α.

It is well-known that $C^\alpha : D(C^\alpha) \subset X \mapsto X$ is a densely defined closed linear operator. Further, its domain $D(C^\alpha)$ is endowed with the norm defined as

$$\|x\|_\alpha = \|C^\alpha x\|, \quad \text{for } x \in D(C^\alpha).$$

Since C is closed, then it can be easily shown that $X_\alpha = (D(C^\alpha), \|.\|_\alpha)$ is also a Banach space.

Recall that if $-C$ is the infinitesimal generator of an analytic semigroup $(R(t))$ and that $0 \in \rho(C)$, for $\alpha > 0$, the fractional powers C^α of C are implicitly defined as

$$C^{-\alpha} = \frac{1}{\Gamma(\alpha)} \int_0^\infty t^{\alpha-1} R(t) dt,$$

where $\Gamma(\alpha)$ is the classical Gamma function.

In the case where $0 < \alpha \leq 1$, since $0 \in \rho(C)$, then the operator $C^{-\alpha}$ is bounded, that is, there exists $K > 0$ such that $\|C^{-\alpha}\| \leq K$.

Theorem 1.21. *Under the above assumptions on the operator C, we have*

(i) $C^{-\alpha} C^{-\beta} = C^{-(\alpha+\beta)}$;

(ii) $\lim_{\alpha \to 0} C^{\alpha} = I$ (strong operator topology).

Proof. See [83] for instance.

We also recall the following.

Lemma 1.22. *Let $-C$ be the infinitesimal generator of an analytic semigroup $R(t)$. Assume that $0 \in \rho(C)$.*

Then for $\alpha > 0$, we have the following:

1. *for every $u \in D(C^{\alpha})$, $\quad R(t)C^{\alpha}u = C^{\alpha}R(t)u$. Moreover $C^{\alpha}R(t)$ is bounded, with an estimate of the form*

$$\|C^{\alpha}R(t)\| \leq M_{\alpha} t^{-\alpha} e^{-\delta t}.$$

2. *If $0 < \alpha \leq 1$ and $u \in D(C^{\alpha})$, we have an estimate of the form*

$$\|R(t)u - u\| \leq C_{\alpha} t^{\alpha} \|C^{\alpha}u\|.$$

More details on fractional powers of operators can be found in the literature, especially in [83].

1.5 Evolution Equations

Unlike the finite dimensional case, the infinite dimensional theory of evolution equations has several notions of solutions. We will present two of them in this section : the so-called classical solutions and mild solutions. We first consider in a Banach space $(X, \|.\|)$ the Abstract Cauchy Problem (ACP)

$$x'(t) = Ax(t) \ (t > 0) \ \ x(0) = x_0 \in D(A) \tag{1.1}$$

where $A : D(A) \subset X \to X$ is a linear operator densely defined in X. We have:

Definition 1.23. *The problem (1.1) is said to be well-posed if $\rho(A) \neq \emptyset$ and for each $x_0 \in D(A)$, there exists a unique (classical) solution $x : [0, \infty) \mapsto D(A)$ in $C^1([0, \infty), X)$.*

It is well-known that "well-posedness" of (1.1) involves more than existence and uniqueness of solutions, and continuous dependence on the initial data. Indeed we have:

Theorem 1.24. *The problem (1.1) above is well-posed iff A generates a C_0-semigroup of linear operators $T = T(t))_{t \in \mathbb{R}^+}$ on X. In this case, the (classical) solution of (1.1) is given by the formula $x(t) = T(t)x_0, \ t \in \mathbb{R}^+$.*

For the proof, refer for instance to [45], pages 83-84.

We now consider the linear non-homogeneous equation

$$x'(t) = Ax(t) + f(t), \ \ t \in \mathbb{R}^+ \tag{1.2}$$

with initial data

$$x(0) = x_0 \in X. \tag{1.3}$$

We assume that A is the infinitesimal genarator of a C_0-semigroup $T = (T(t))_{t \in \mathbb{R}^+}$, and $f \in L^1_{loc}(\mathbb{R}^+, X)$. Then we have:

Definition 1.25. *A function* $x : \mathbb{R}^+ \mapsto X$ *is said to be a mild solution of the Abstract Cauchy Problem (1.2)-(1.3) if* $x \in C(\mathbb{R}^+, X)$ *and satisfies the equation*

$$x(t) = T(t)x(a) + \int_a^t T(t-s)f(s)ds, \qquad (1.4)$$

for any $a \in \mathbb{R}$, *and any* $t \geq a$.

This formula is also called the *Variation of Constants Formula.* We also have:

Definition 1.26. *If* $f \in C(\mathbb{R}^+, X)$, *then a function* $x : \mathbb{R}^+ \mapsto X$ *is said to be a classical solution of the Problem (1.2)-(1.3) if:*

(i) $x \in C(\mathbb{R}^+, X)$ *and* $x(0) = x_0$
(ii) x *is strongly differentiable in* X *at each* $t \in \mathbb{R}^+$
(iii) x *satisfies the equation (1.2) in* X *everywhere on* $(0, \infty)$
(iv) $x'(t) \in L^1_{loc}(\mathbb{R}^+, X)$ *and satisfies*

$$x(t) = x(a) + \int_a^t x'(s)ds, \quad \text{for all} \ \ a \in \mathbb{R}, \ \ \text{all} \ \ t \geq a.$$

It is clear that classical solutions are mild solutions too. The converse is not always true.

We have the following (see [45, p. 84]):

Theorem 1.27. *Suppose* A *generates a* C_0-*semigroup on* X *and* $x_0 \in D(A)$. *Assume either*

(i) $f \in C(\mathbb{R}^+, X)$ *takes values in* $D(A)$ *and* $Af \in C(\mathbb{R}^+, X)$, *or*
(ii) $f \in C^1(\mathbb{R}^+, X)$.

Then the Abstract Cauchy Problem (1.2)-(1.3) has a unique (classical) solution with $x : [0, \infty) \mapsto D(A)$ *such that* $x(0) = x_0$.

Theorem 1.28. *Assume that A generates a C_0-semigroup on X and $f \in L^1_{loc}(\mathbb{R}^+, X)$.*

Then a mild solution $x : [0, \infty) \mapsto X$ of the (ACP) is a classical solution if and only if both of the following conditions are satisfied:

(i) $x(t) \in D(A)$ almost everywhere on $[0, \infty)$, and
(ii) $Ax \in L^1_{loc}(\mathbb{R}^+, X)$, i.e. $x \in L^1_{loc}(\ R^+, D(A))$.

See for instance [87, p. 149] for the proof.

1.6 Almost Automorphic Functions

Definition 1.29. *Let $(X, \|.\|)$ be a (real or complex) Banach space and $f : \mathbb{R} \to X$ a (strongly) continuous function. We say that f is almost automorphic if for every sequence of real numbers (s'_n), there exists a subsequence (s_n) such that:*

$$\lim_{m \to \infty} \lim_{n \to \infty} f(t + s_n - s_m) = f(t)$$

for each $t \in \mathbb{R}$.

This limit means that:

$$g(t) := \lim_{n \to \infty} f(t + s_n)$$

is well defined for each $t \in \mathbb{R}$ and

$$\lim_{n \to \infty} g(t - s_n) = f(t)$$

for each $t \in \mathbb{R}$.

Remark 1.30. We observe that the function $g(t)$ in *Definition 1.29* is measurable but not necessarily continuous.

If the convergence above is uniform on $\mathbb{R}$, the function f is said to be almost periodic (in Bochner's sense).

Also if the convergence above holds in the weak topology of X, that is

$$g(t) := w - \lim_{n\to\infty} f(t + s_n)$$

is well defined for each $t \in \mathbb{R}$ and

$$w - \lim_{n\to\infty} g(t - s_n) = f(t)$$

for each $t \in \mathbb{R}$,

we say that f is weakly almost automorphic.

Clearly almost automorphy implies weak-almost automorphy. The reader can find more informations on weak almost automorphy in [80].

We end these remarks by the following important result (see *Theorem 2.1.10* in [80]):

If $f_n : \mathbb{R} \mapsto X$, $n = 1, 2...$, is a sequence of almost automorphic functions such that $\lim_{n\to\infty} f_n(t) = f(t)$, uniformly in $t \in \mathbb{R}$, then f is also almost automorphic.

Theorem 1.31. *If f, $f_1, f_2 : \mathbb{R} \to X$ are almost automorphic functions , then the following are true:*

i) $f_1 + f_2$ is almost automorphic .

ii) cf is almost automorphic for every scalar c.

iii) $f_a(t) \equiv f(t + a)$ is almost automorphic for each fixed $a \in \mathbb{R}$.

iv) $\sup_{t\in\mathbb{R}} \|f(t)\| < \infty$, that is f is a bounded function.

v) The range $R_f = \{f(t) : t \in \mathbb{R}\}$ of f is relatively compact in
   X.

Proof: See [80] .   □

Proposition 1.32. *We have*

(i) $\sup_{t \in \mathbb{R}} \|g(t)\| = \sup_{t \in \mathbb{R}} \|f(t)\|$
(ii) $R_g \subseteq \overline{R_f}$, *where g is the function that appears in Definition*
   1.29.

Let us first recall the following.

Lemma 1.33. *(see [90])*

(i) Weakly bounded sets are bounded (in norm) in any Banach
   space $(X, \|.\|)$, and
(ii) if $w\text{-}\lim_{n \to \infty} x_n = x$, then $\|x\| \leq \liminf_{n \to \infty} \|x_n\|$, and the
   sequence $(\|x_n\|)$ is bounded.

Now we prove *Proposition 1.32.*
Proof. (i) Since $g(t) = \lim_{n \to \infty} f(t + s_n)$, we may use the above
Lemma to get, for each $t \in \mathbb{R}$, the inequality

$$\|g(t)\| \leq \liminf_{n \to \infty} \|f(t + s_n)\|$$

so that

$$\sup_{t \in \mathbb{R}} \|g(t)\| \leq \sup_{t \in \mathbb{R}} \|f(t)\|.$$

The reverse inequality is proved the same way.

(ii) is straightforward.   □

We also establish the following composition result:

Theorem 1.34. *Let X, Y be two Banach spaces with norms $\|.\|_X$ and $\|.\|_Y$ respectively, and let $f : \mathbb{R} \to X$ be an almost automorphic function. If $\phi : X \to Y$ is a continuous function, then the composite function $\phi(f(t)) : \mathbb{R} \to Y$ is almost automorphic.*

Proof: Since f is almost automorphic, $\overline{f(\mathbb{R})}$ is compact in X by *Theorem 1.31 v)*. We deduce that ϕ restricted to $\overline{f(\mathbb{R})}$ is uniformly continuous. So given $\varepsilon > 0$, there exists $\delta > 0$ such that

$$\|\phi(x_1) - \phi(x_2)\|_Y < \varepsilon, \quad \text{for any} \ \ x_1, x_2 \in \overline{f(\mathbb{R})}$$

with $\|x_1 - x_2\|_X < \delta$.

Let (s'_n) be an arbitrary sequence of real numbers. There exists a subsequence $(s_n) \subset (s'_n)$ such that

$$g(t) := \lim_{n \to \infty} f(t + s_n)$$

is well-defined for each $t \in \mathbb{R}$ and

$$\lim_{n \to \infty} g(t - s_n) = f(t)$$

for each $t \in \mathbb{R}$.

Note that ϕ is well-defined on $g(\mathbb{R})$, since $g(\mathbb{R}) \subset \overline{f(\mathbb{R})}$ *(Proposition 1.32 ii))*. Therefore there exists $N \in \mathbb{N}$ such that if $n > N$, we have

$$\|f(t + s_n) - g(t)\|_X < \delta$$

for each $t \in \mathbb{R}$.

Thus

$$\|\phi(f(t + s_n)) - \phi(g(t))\|_Y < \varepsilon$$

for each $t \in \mathbb{R}$; which shows that

$$\lim_{n\to\infty} \phi(f(t+s_n)) = \phi(g(t))$$

for each $t \in \mathbb{R}$.

Analogously, we can prove that

$$\lim_{n\to\infty} \phi(g(t-s_n)) = \phi(f(t))$$

for each $t \in \mathbb{R}$. So that $t \to \phi(f(t))$ is almost automorphic. The proof is now complete. $\square$

Proposition 1.35. *Let $\phi : \mathbb{R} \to \mathbb{C}$ and $f : \mathbb{R} \to X$ be almost automorphic where X is a complex Banach space. Then $\phi f : \mathbb{R} \to X$ defined by $(\phi f)(t) = \phi(t)f(t)$ is also almost automorphic.*

Proof. Both functions ϕ and f are bounded since they are almost automorphic (*Theorem 1.31 iv)*). So we let $K_1 = \sup_{t\in\mathbb{R}} |\phi(t)|$.

Let (t_n) be an arbitrary sequence of real numbers. Then there exists a subsequence (s_n) of (t_n) such that:

$$v(t) := \lim_{n\to\infty} \phi(t+s_n)$$

is well-defined for each $t \in \mathbb{R}$, and

$$\lim_{n\to\infty} v(t-s_n) = \phi(t)$$

for each $t \in \mathbb{R}$.

Also

$$g(t) := \lim_{n\to\infty} f(t+s_n)$$

is well-defined for each $t \in \mathbb{R}$, and

$$\lim_{n\to\infty} g(t - s_n) = f(t)$$

for each $t \in \mathbb{R}$.

Now write

$$\phi(t + s_n)f(t + s_n) - v(t)g(t) = \phi(t + s_n)f(t + s_n) - \phi(t + s_n)g(t)$$
$$+ \phi(t + s_n)g(t) - v(t)g(t).$$

So that

$$\|\phi(t + s_n)f(t + s_n) - v(t)g(t)\| \le K_1\|f(t + s_n) - g(t)\|$$
$$+ K_2|\phi(t + s_n) - v(t)|,$$

where $K_2 = \sup_{t\in\mathbb{R}} \|g(t)\| = \sup_{t\in\mathbb{R}} \|f(t)\| < \infty$ (*Proposition 1.32 i*) and *Theorem 1.31 iv*)).

Clearly when $n \to \infty$, we obtain

$$\lim_{n\to\infty} \phi(t + s_n)f(t + s_n) = v(t)g(t)$$

for each $t \in \mathbb{R}$.

It is also easy to check that

$$\lim_{n\to\infty} v(t - s_n)g(t - s_n) = \phi(t)f(t)$$

for each $t \in \mathbb{R}$.

The proof is now complete.   □

The following result is important in view of its applications to the theory of evolution equations (see for instance *Chapter 2*, [80]).

Theorem 1.36. *Let* $T = (T(t))_{t\in\mathbb{R}}$ *be a one parameter group of strongly continuous linear operators such that* $\sup_{t\in\mathbb{R}} \|T(t)\| =$

$M < \infty$. Let $f : \mathbb{R} \to X$ be an almost automorphic function and $S = f(\mathbb{Q})$, where $\mathbb{Q}$ denotes the set of rational numbers, with the property that the function $T(t)x : \mathbb{R} \to X$ is almost automorphic for each $x \in S$.

Then $T(t)f(t) : \mathbb{R} \to X$ is almost automorphic.

Proof: Let $B = \{f(t) : t \in \mathbb{R}\}$ be the range of f. Then S is a countable subset of B. It is also dense in $\overline{B}$, the closure of B. Indeed it is known that if g is a continuous function $g : X_1 \to X_2$ where X_1 and X_2 are two topological spaces and $A \subset X_1$, then

$$g(\overline{A}) \subset \overline{g(A)}.$$

Since f is continuous and $\overline{\mathbb{Q}} = \mathbb{R}$, then we have

$$B = f(\mathbb{R}) = f(\overline{\mathbb{Q}}) \subset \overline{f(\mathbb{Q})} = \overline{S}$$

That is $\overline{S} = \overline{B}$ since S is a subset of B; which proves our claim.

Let $S = (x_n)$; then $T(t)x_n$ is almost automorphic for each $n = 1, 2, \dots$. Consider an arbitrary sequence of real numbers (s'_n). Using the well known Cantor diagonal procedure we can show that there exists a subsequence (s_n) of (s'_n) such that

$$\lim_{n \to \infty} T(s_n)x \ \text{ exists for every } \ x \in S.$$

Pick x_0 arbitrary in $\overline{B}$. For any n, m, k we have

$$\|T(s_n)x_0 - T(s_m)x_0\| \leq \|T(s_n)x_0 - T(s_n)x_k\|$$
$$+ \|T(s_n)x_k - T(s_m)x_k\|$$
$$+ \|T(s_m)x_k - T(s_m)x_0\|$$

$$\leq \|T(s_n)\|\|x_0 - x_k\|$$
$$+ \|T(s_n)x_k - T(s_m)x_k\|$$
$$+ \|T(s_m)\|\|x_k - x_0\|$$
$$\leq 2M\|x_0 - x_k\| + \|T(s_n)x_k - T(s_m)x_k\|.$$

Now, since $x_k \in S$, $(T(s_i)x_k)_{i\in\mathbb{N}}$ is a Cauchy sequence because it is convergent as seen above.

Thus

$$\lim_{n,m\to\infty} \|T(s_n)x_k - T(s_m)x_k\| = 0$$

and consequently:

$$\lim_{n,m\to\infty} \|T(s_n)x_0 - T(s_m)x_0\| \leq 2M\|x_0 - x_k\|.$$

Therefore, using the fact that S is dense in $\overline{B}$, we can say that

$$\lim_{n\to\infty} T(s_n)x_0 \text{ exists for every } x_0 \in \overline{B}.$$

Now we observe that $\lim_{n\to\infty} T(s_n)x = y$ defines a mapping F from the linear subspace spanned by $\overline{B}$ into X, namely

$$Fx = y \text{ if } \lim_{n\to\infty} T(s_n)x = y.$$

The map F has the following properties:

i) F is linear,

ii) $\|Fx\| = \|y\| \leq \lim_{n\to\infty} \|T(s_n)x\| \leq M\|x\|$ for every x in the subspace spanned by $\overline{B}$,

iii) F is one-to-one,

iv) If (x_n) is a given sequence in $\overline{B}$ such that strong-$\lim_{n\to\infty} x_n = x$ exists, then

strong-$\lim_{n\to\infty} T(s_n)x_n = Fx$, and strong-$\lim_{n\to\infty} Fx_n = Fx$.

Let $R_F = \{Fx : \ x \in \overline{B}\}$ be the range of F. Then we observe that $\overline{F(S)} = R_F$ and R_F is compact in X. Let us show that

$$\lim_{n \to \infty} T(-s_n)y \ \text{ exists for every } \ y \in R_F.$$

It suffices to prove that

$$\lim_{n \to \infty} T(-s_n)y_m \ \text{ exists for every } \ y_m \in F(S).$$

where $y_m = Fx_m, \ m = 1, 2, \dots$.

Since $T(t)x_m$ is almost automorphic for each $m = 1, 2, \dots$ we have

$$\begin{aligned}
\lim_{n \to \infty} T(t + s_n)x_m &= \lim_{n \to \infty} T(t)T(s_n)x_m \\
&= T(t) \lim_{n \to \infty} T(s_n)x_m \\
&= T(t)Fx_m \\
&= T(t)y_m
\end{aligned}$$

pointwise on $\mathbb{R}$. Also we have

$$\begin{aligned}
\lim_{n \to \infty} T(t - s_n)y_m &= T(t)x_m \\
&= T(t) \lim_{n \to \infty} T(-s_n)y_m.
\end{aligned}$$

Now, for $t = 0$, we get

$$\lim_{n \to \infty} T(-s_n)y_m \ \text{ exists for } \ m = 1, 2, \dots$$

and $T(0)x_m = x_m$. Hence, we have

$$\lim_{n \to \infty} T(-s_n)y \ \text{ exists for every } \ y \in R_F.$$

This defines a linear function G on the linear subspace spanned by R_F where

$$Gy = \lim_{n\to\infty} T(-s_n)y.$$

It is easy to verify that G has the same properties as did F and we observe that

$$GFx = x \quad \text{for every} \quad x \in \overline{B}.$$

Let (s'_n) be an arbitrary sequence of real numbers. Then we can extract a subsequence (s_n) such that

$$\lim_{n\to\infty} f(t + s_n) = g(t)$$

and

$$\lim_{n\to\infty} g(t - s_n) = f(t)$$

pointwise on $\mathbb{R}$, and

$$\lim_{n\to\infty} T(-s_n)x = y \quad \text{exists for each} \quad x \in \overline{B}.$$

Now let us observe that for every $t \in \mathbb{R}$ and $n = 1, 2, \ldots$ we have

$$f(t + s_n), \quad g(t) \in \overline{B}.$$

Let t be arbitrary in $\mathbb{R}$. Then for every $n = 1, 2, \ldots$, we get

$$T(t + s_n)f(t + s_n) = T(t)T(s_n)f(t + s_n)$$

so that

$$\lim_{n\to\infty} T(t + s_n)f(t + s_n) = T(t)Fg(t)$$

and

$$\lim_{n\to\infty} T(t - s_n)Fg(t - s_n) = T(t)\lim_{n\to\infty} T(-s_n)Fg(t - s_n)$$
$$= T(t)GFf(t)$$
$$= T(t)f(t),$$

since $f(t) \in \overline{B}$. The theorem is proved. $\square$

Theorem 1.37. *Assume that $(T(t))_{t \geq 0}$ is a C_0-group of bounded linear operators on X and let $x(t) = T(t)x_0$ is almost automorphic for some $x_0 \in X$.*

Then

$$\inf_{t \in \mathbb{R}} \|x(t)\| > 0, \quad or \quad x(t) = 0 \quad for \ every \ t \in \mathbb{R}.$$

Proof: Assume that $\inf_{t \in \mathbb{R}} \|x(t)\| = 0$ and let (s_n') be a minimizing sequence of real numbers, that is $\lim_{n \to \infty} \|x(s_n')\| = 0$. We can extract a subsequence $(s_n) \subseteq (s_n')$ such that

$$y(t) := \lim_{n \to \infty} x(t + s_n)$$

is well defined for each $t \in \mathbb{R}$, and

$$\lim_{n \to \infty} y(t - s_n) = x(t)$$

for each $t \in \mathbb{R}$.

We also have

$$x(t + s_n) = T(t + s_n)x_0 = T(t)T(s_n)x_0 = T(t)x(s_n).$$

Thus

$$y(t) = \lim_{n \to \infty} x(t + s_n) = T(t) \lim_{n \to \infty} x(s_n) = 0,$$

for each $t \in \mathbb{R}$; which shows that $y(t) = 0$ for each $t \in \mathbb{R}$ and consequently $x(t) = 0$ identically on $\mathbb{R}$. $\square$

Differentiability and integration of almost automorphic functions are presented in [80]. We recall the following important Bohr-Amerio type result:

If $f : \mathbb{R} \to X$ is almost aumorphic and $F : \mathbb{R} \to X$ defined by $F(t) := \int_0^t f(t)dt$ has a relatively compact range in X, then F is also almost automorphic.

In the case X is a uniformly convex Banach space, the conclusion holds true if the range of F is bounded in X.

In conclusion to this section, let us note that the set $AA(X)$ of all almost automorphic functions $\mathbb{R} \to X$, $((X, \| \cdot \|)$ a Banach space), is a linear vector space in view of *Theorem 1.31.*

Equipped with the norm

$$\|f\|_{AA(X)} = \sup_{t \in \mathbb{R}} \|f(t)\|,$$

$AA(X)$ turns out to be a Banach space in view of the above remarks.

1.6.1 Asymptotically Almost Automorphic Functions

Definition 1.38. *Let $(X, \| \cdot \|)$ be a (real or complex) Banach space. A continuous function $f : \mathbb{R}^+ \to X$ is said to be asymptotically almost automorphic if it admits a decomposition*

$$f(t) = g(t) + h(t), \quad t \in \mathbb{R}^+$$

where $g : \mathbb{R} \to X$ is an almost automorphic function, $h : \mathbb{R}^+ \to X$ is a continuous function with $\lim_{t \to \infty} \|h(t)\| = 0$.

g and h are called respectively the principal and corrective terms of the function f.

We have the following immediate facts (see [80] for details):

Theorem 1.39. *If f, f_1, f_2 are asymptotically almost automorphic, then $f_1 + f_2$ and λf, λ an arbitrary scalar, are also asymptotically almost automorphic.*

We also have the important result:

Theorem 1.40. *The decomposition of an asymptotically almost automorphic function is unique.*

Denote $AAA(X)$ the linear vector space of all asymptotically almost automorphic functions $f : \mathbb{R}^+ \mapsto X$. It is clear that the formula:

$$\|f\|_{AAA(X)} = \|g(t)\|_{AA(X)} + \sup_{t \in \mathbb{R}_+} \|h(t)\| \qquad (1.5)$$

where g and h are the principal and corrective terms of f, respectively, defines a norm on the space $AAA(X)$.

The following holds true.

Theorem 1.41. $AAA(X)$ *is a Banach space.*

Proof: Let (f_n) be a Cauchy sequence in $AAA(X)$, with (g_n) and (h_n) as respective principal and corrective terms. It is clear that (g_n) is a Cauchy sequence in the Banach space of all almost automorphic functions $AA(X)$. Thus there exists $g \in AA(X)$ such that $g_n \to g$ uniformly on $\mathbb{R}$.

Moreover the corrective terms (h_n) also form a Cauchy sequence of continuous functions with respect to the norm sup.

We then deduce that there exists a function $h \in C(\mathbb{R}_+, X)$, such that $h_n \mapsto h$ uniformly on $\mathbb{R}_+$. Using the fact that for each

$n = 1, 2, ...$, $\lim_{t \to \infty} \|h_n(t)\| = 0$, and the equality $h(t) = (h(t) - h_n(t)) + h_n(t)$ for $t \in \mathbb{R}_+$, we obtain

$$\lim_{t \to \infty} \|h(t)\| = 0.$$

This implies that the function f defined as $f := g + h \in AAA(X)$ and $\lim_{n \to \infty} \|f_n - f\| = 0$, thus $AAA(X)$ is a Banach space. $\square$

1.6.2 Applications to Abstract Dynamical Systems

In this section, we will study the behavior of asymptotically almost automorphic semigroups of linear operators $T = (T(t))_{t \in \mathbb{R}^+}$ as t tends to infinity. We will present some topological and asymptotic properties based on the classical Nemytskii-Stepanov theory of dynamical systems.

First of all we present a connection between the so-called abstract dynamical systems and C_0-semigroups of linear operators. $(X, \| \cdot \|)$ will denote a Banach space (over $\mathbb{R}$ or $\mathbb{C}$).

Definition 1.42. *A mapping $u : \mathbb{R}^+ \times X \to X$ is called an (abstract) dynamical system if*

i) $u(0, x) = x$, for every $x \in X$.
ii) $u(\cdot, x) : \mathbb{R}^+ \to X$ is continuous for any $t > 0$ and right-continuous at $t = 0$, for each $x \in X$.
iii) $u(t, \cdot) : X \to X$ is continuous for each $t \in \mathbb{R}^+$.
iv) $u(t + s, x) = u(t, u(s, x))$, for all $t, s \in \mathbb{R}^+$ and $x \in X$.

If $u : \mathbb{R}^+ \times X$ is a dynamical system, the mapping $u(\cdot, x) : \mathbb{R}^+ \to X$ will be called a motion originating at $x \in X$.

Now we are ready to state and prove the following basic result:

Theorem 1.43. *Every C_0-semigroup $(T(t))_{t \in \mathbb{R}^+}$ determines a dynamical system and conversely by defining $u(t, x) = T(t)x$, $t \in \mathbb{R}^+$, $x \in X$.*

Proof: Let $u(t, x)$ be a dynamical system in the sense of *Definition 1.42* above and consider

$$T(t)x = u(t, x), \quad t \in \mathbb{R}^+, \ x \in X.$$

Then obviously $T(0) = I$, the identity operator on X since for every $x \in X$, $T(0)x = u(0, x) = x$.

Let $t, s \in \mathbb{R}^+$ and $x \in X$; then we have

$$T(t + s)x = u(t, s, x) = u(t, u(s, x))$$

by property iv) of *Definition 1.42*. But we have also

$$T(t)T(s)x = T(t)u(s, x) = T(t, u(s, x))$$

using the definition of $T(t)x$. Therefore,

$$T(t + s)x = T(t)T(s)x,$$

for every $t, s \in \mathbb{R}^+$, $x \in X$, which proves the semigroup property

$$T(t + s)x = T(t)T(s)x,$$

for all $t, s \in \mathbb{R}^+$.

Continuity of $T(t)x : X \to X$ follows readily from property iii) of *Definition 1.42*, for every $t \in \mathbb{R}^+$.

Now we have

$$\lim_{t \to 0^+} T(t)x = \lim_{t \to 0^+} u(t, x) = u(0, x) = x$$

using properties ii) then i) in the above *Definition 1.42*. We have proved that $(T(t))_{t \in \mathbb{R}^+}$ is a C_0-semigroup.

Conversely, suppose we have a C_0-semigroup $(T(t))_{t \in \mathbb{R}^+}$ and define $u : \mathbb{R}^+ \times X \to X$ by

$$u(t, x) = T(t)x, \quad t \in \mathbb{R}^+, \ x \in X.$$

Then all properties i)-iv) in *Definition 1.42* are obviously true. The mapping u is then a dynamical system. □

Theorem 1.43 tells us that the notions of abstract dynamical systems and C_0-semigroups are equivalent. This fact provides a solid ground to study C_0-semigroups of linear operators as an independent topic.

In the rest of the section, we will consider a C_0-semigroup of linear operators $T = (T(t))_{t \in \mathbb{R}^+}$ such that the motion

$$T(t)x_0 : \mathbb{R}^+ \to X$$

is in $AAA(X)$ with principal term $f(t)$.

Let us now introduce some notations and definitions.

We let x_0 be some fixed element of X.

Definition 1.44. *A function $\varphi : \mathbb{R} \to X$ is said to be a complete trajectory of T if it satisfies the functional equation*

$$\varphi(t) = T(t - a)\varphi(a),$$

for all $a \in \mathbb{R}$ and all $t \geq a$.

We have also the following properties.

Theorem 1.45. *The principal term of $T(t)x_0$ is a complete trajectory for T.*

Proof: We have $T(t)x_0 = f(t) + h(t)$, $t \in \mathbb{R}^+$. Since f is almost automorphic, there exists a subsequence $(n_k) \subseteq (n) = \mathbb{N}$ such that

$$g(t) := \lim_{k \to \infty} f(t + n_k)$$

is well-defined for each $t \in \mathbb{R}$ and

$$\lim_{k \to \infty} g(t - n_k) = f(t)$$

pointwise on $\mathbb{R}$.

Put $\varphi(t) = T(t)x_0$. Then $\varphi(0) = x_0$. Let us fix $a \in \mathbb{R}$ and choose k large enough so that $a + n_k \geq 0$. If $s \geq 0$, then

$$\begin{aligned}
\varphi(a + s + n_k) &= T(a + s + n_k)\varphi(0) \\
&= T(s)T(a + n_k)\varphi(0) \\
&= T(s)\varphi(a + n_k).
\end{aligned}$$

Consequently,

$$f(a + s + n_k) + h(a + s + n_k) = T(s)\varphi(a + n_k)$$

where $s \geq 0$ and $a + n_k \geq 0$.

But we have

$$\lim_{k \to \infty} f(a + s + n_k) = g(a + s), \ \lim_{k \to \infty} h(a + s + n_k) = 0,$$

so

$$\lim_{k\to\infty} \varphi(a + s + n_k) = \lim_{k\to\infty} T(s)\varphi(a + n_k) = g(a + s).$$

We also have

$$\lim_{k\to\infty} \varphi(a + n_k) = g(a).$$

Using continuity of $T(t)$, we get

$$\lim_{k\to\infty} T(s)\varphi(a + n_k) = T(s)g(a).$$

We can now establish the following equality

$$T(s)g(a) = g(a + s), \quad \forall a \in \mathbb{R}, \ \forall s \geq 0.$$

But we have

$$\lim_{k\to\infty} g(t - n_k) = f(t) , \quad \text{for each} \quad t \in \mathbb{R}$$

and

$$g(a - n_k + s) = T(s)g(a - n_k) , \quad \forall a \in \mathbb{R}, \quad \forall s \geq 0.$$

Therefore

$$\lim_{k\to\infty} g(a - n_k + s) = T(s)f(a) , \quad \forall a \in \mathbb{R}, \quad \forall s \geq 0$$

so that

$$f(a + s) = T(s)f(a) , \quad \forall a \in \mathbb{R}, \quad \forall s \geq 0.$$

Finally let us put $s = t - a$ with $t \geq 0$. Then

$$f(t) = T(t - a)f(a), \quad \forall a \in \mathbb{R}, \quad \forall t \geq a.$$

The proof is complete. □

Definition 1.46. *The*

$$\omega^+(x_0) = \{y \in X \, / \, \exists \, 0 \le t_n \to \infty : \lim_{n \to \infty} T(t_n)x_0 = y\},$$

is called the ω-limit set of $T(t)x_0$.

$$\omega_f^+(x_0) = \{y \in X \, / \, \exists \, 0 \le t_n \to \infty : \lim_{n \to \infty} f(t_n) = y\}$$

is called the ω-limit set of $f(t)$, the principal term of $T(t)x_0$.

$$\gamma^+(x_0) = \{T(t)x_0 \, / \, t \in \mathbb{R}^+\}$$

is the trajectory of $T(t)x_0$.

We have the following properties.

Theorem 1.47. $\omega^+(x_0) \ne \emptyset$.

Proof: We let $t_n = n$, $n = 1, 2, \cdots$. Since $f \in AA(X)$, there exists a subsequence $(t_{n_k}) \subset (t_n)$ with $t_{n_k} = n_k$ such that

$$\lim_{k \to \infty} f(t_{n_k}) = g(0).$$

But

$$\lim_{k \to \infty} T(t_{n_k})x_0 = \lim_{k \to \infty} f(t_{n_k}).$$

We then get

$$\lim_{k \to \infty} T(t_{n_k})x_0 = g(0).$$

Consequently, $g(0) \in \omega^+(x_0)$, since $t_{n_k} \to \infty$ as $k \to \infty$. So $\omega^+(x_0)$ is not empty.

The proof is complete. $\square$

Theorem 1.48. $\omega^+(x_0) = \omega_f^+(x_0)$.

Proof: To see that $T(t)x_0$ and its principal term have the same ω-limit set, it suffices to observe that

$$\lim_{t \to \infty} T(t)x_0 = \lim_{t \to \infty} f(t).$$

The proof is complete. □

Definition 1.49. *A subset $B \subseteq X$ is said to be invariant set under the semigroup $T = (T(t))_{t \in \mathbb{R}^+}$ if $T(t)y \in B$ for every $y \in B$ and $t \in \mathbb{R}^+$.*

Theorem 1.50. $\omega^+(x_0)$ *is invariant under T.*

Proof: Let $y \in \omega^+(x_0)$; then there exists $0 \leq t_n \to \infty$ such that $\lim_{t \to \infty} T(t_n)x_0 = y$.

Consider the sequence (s_n) where $s_n = t + t_n, n = 1, 2, \cdots$ for a given $t \in \mathbb{R}^+$. Then $s_n \to \infty$ as $n \to \infty$. We have

$$T(s_n)x_0 = T(t)T(t_n)x_0, \quad n = 1, 2 \cdots$$

and $\lim_{n \to \infty} T(s_n)x_0 = T(t)y$, using continuity of $T(t)$. Therefore $T(t)y \in \omega^+(x_0)$.

This completes the proof. □

Theorem 1.51. $\omega^+(x_0)$ *is closed in X.*

Proof: Let $y \in \overline{\omega^+(x_0)}$ be the closure of $\omega^+(x_0)$; then there exists a sequence of elements $y_m \in \omega^+(x_0), m = 1, 2, \ldots$ such that $y_m \to y$. For each y_m, there exists $0 \leq t_{m,n} \to +\infty$, as $n \to +\infty$ such that $\lim_{n \to \infty} T(t_{m,n})x_0 = y_m$. Recursively choose

$t_{1,n_1} > 1$ such that $\|y_1 - T(t_{1,n_1})x_0\| < \frac{1}{2}$

$t_{2,n_2} > \max(2, t_{1,n_1})$ such that $\|y_2 - T(t_{2,n_2})x_0\| < \frac{1}{2^2}$

$t_{3,n_3} > \max(3, t_{2,n_2})$ such that $\|y_3 - T(t_{3,n_3})x_0\| < \frac{1}{2^3}$

$t_{k,n_k} > \max(k, t_{k-1,n_{k-1}})$ such that $\|y_k - T(t_{k,n_k})x_0\| < \frac{1}{2^k}$.

Let $s_k = t_{k,n_k}, k = 1, 2, \cdots$. Clearly $0 < s_k \to +\infty$ as $k \to +\infty$, and we have

$$\|T(s_k)x_0 - y\| \le \|T(s_k)x_0 - y_k\| + \|y_k - y\|$$
$$< \frac{1}{2^k} + \|y_k - y\|.$$

Since $\lim_{k \to +\infty} y_k = y$, we have $y \in \omega^+(x_0)$.

This achieves the proof.   □

Theorem 1.52. $\omega^+(x_0)$ *is compact if* $\gamma^+(x_0)$ *is relatively compact.*

Proof: It is obvious that $\omega^+(x_0) \subset \overline{\gamma^+(x_0)}$ the closure of $\gamma^+(x_0)$. But $\overline{\gamma^+(x_0)}$ is a compact set and $\omega^+(x_0)$ is a closed set (see *Theorem 1.51*). Therefore $\omega^+(x_0)$ is itself compact.   □

Theorem 1.53. $\gamma_f(x_0) = \{f(t) \,/\, t \in \mathbb{R}\}$ *is invariant under the semigroup* T.

We recall also that $\gamma_f(x_0)$ *is relatively compact, since* $f(t)$ *is almost automorphic.*

Proof: Let $y \in \gamma_f(x_0)$. So there exists $\sigma \in \mathbb{R}$ such that $y = f(\sigma)$. For arbitrary $a \in \mathbb{R}$ such that $\sigma \ge a$, we can write

$$y = f(\sigma) = T(\sigma - a)f(a),$$

since f is a complete trajectory *Theorem 1.45*. Now let $t \geq 0$. Then

$$T(t)y = T(t + \sigma - a)f(a)$$
$$= f(t + \sigma),$$

i.e., $T(t)y \in \gamma_f(x_0)$, $\forall t \geq 0$.

$\gamma_f(x_0)$ is indeed invariant under the semigroup T.   □

Theorem 1.54. *Let* $\nu(t) = \inf_{y \in \omega^+(x_0)} \|T(t)x_0 - y\|$.
Then

$$\lim_{t \to +\infty} \nu(t) = 0.$$

Proof: Suppose not, that is $\lim_{t \to +\infty} \nu(t) \neq 0$. Then there exists $\varepsilon > 0$ such that for every $n = 1, 2, \cdots$ there exists $t'_n \geq n$ such that $\nu(t'_n) \geq \varepsilon$, i.e.,

$$\exists t'_n \geq n, \ \|T(t'_n)x_0 - y\| \geq \varepsilon \quad \forall y \in \omega^+(x_0), \ \forall n = 1, 2, \cdots.$$

Let $(t_n)_{n=1}^\infty$ be a subsequence of $(t'_n)_{n=1}^\infty$ such that $(f(t_n))$ converges, say to $\bar{y}$, as is guaranteed by the relative compactness of $\gamma_f(x_0)$.

Now since $t_n \to \infty$ as $n \to \infty$, we get

$$\lim_{n \to \infty} T(t_n)x_0 = \lim_{n \to \infty} f(t_n) = \bar{y}.$$

Therefore $\bar{y} \in \omega^+(x_0)$, which is a contradiction.   □

Remark 1.55. This minimality property shows that the ω-limit set $\omega^+(x_0)$ is the smallest closed set towards which the asymptotically almost automorphic function $T(t)x_0$ tends as t goes to infinity.

Definition 1.56. *$e \in X$ is called a rest-point for the semigroup T if $T(t)e = e$, $\forall t \geq 0$.*

Theorem 1.57. *If x_0 is a rest-point of the semigroup T, then $\omega^+(x_0) = \{x_0\}$.*

Proof: Since $T(t)x_0 = x_0$, $\forall t \geq 0$, then for every sequence of real numbers $(t_n)_{n=1}^{\infty}$ such that $0 \leq t_n \to +\infty$, we get

$$\lim_{n \to +\infty} T(t_n)x_0 = x_0,$$

i.e., $x_0 \in \omega^+(x_0)$.

Now let be $y \in \omega^+(x_0)$. There exists $0 \leq s_n \to \infty$, such that $\lim_{n \to \infty} T(s_n)x_0 = y$. But $T(s_n)x_0 = x_0, n = 1, 2, \cdots$. Therefore $x_0 = y$.

The proof is now complete. $\square$

1.7 Almost Periodic Functions

Definition 1.58. *Let $E = E(\tau)$ be a complete Hausdorff locally convex space. A continuous function $f : \mathbb{R} \to E$ is said to be almost periodic if for each neighborhood of the origin U there exists a real number $l > 0$ such that every interval $[a, a + l]$ contains at least one point s such that*

$$f(t + s) - f(t) \in U, \quad \text{for every} \quad t \in \mathbb{R}.$$

The numbers s depend on U and are called U-translation numbers or U-almost periods of the function f.

Remark 1.59. From *Definition 1.58*, we observe that for each neighborhood of the origin U, the set of U-translation numbers is relatively dense in $\mathbb{R}$.

Theorem 1.60. *(i) If $f : \mathbb{R} \mapsto E$ is an almost periodic function, then f is uniformly continuous.*

(ii) If (f_n) is a sequence of almost periodic functions, $f_n : \mathbb{R} \mapsto E$, $n = 1, 2, 3, \ldots$ such that (f_n) converges uniformly to f on $\mathbb{R}$, then f is also almost periodic.

The following Criterion due to Bochner is a key result.

Theorem 1.61. (Bochner's Criterion). *Let E be a Fréchet space, that is a Hausdorff locally convex space whose topology is induced by a complete and invariant metric.*

Then $f \in C(\mathbb{R}, E)$ is almost periodic if and only if for every sequence of real numbers (s'_n), there exists a subsequence (s_n) such that $(f(t + s_n))$ converges uniformly in $t \in \mathbb{R}$.

Now we denote $AP(E)$ the set of all almost periodic functions $\mathbb{R} \to E$, where E is a Fréchet space. By *Theorem 1.60* and *Theorem 3.1.3* and *Theorem 3.1.9 i)* in [80], $AP(E)$ is a linear space.

We also have the following result (see [16] for details).

Theorem 1.62. $AP(E)$ *is a Fréchet space.*

Proof: Denote by $C(\mathbb{R}, E)$ the linear space of all continuous bounded functions $\mathbb{R} \to E$ and by (q_n), $n \in \mathbb{N}$, the family of seminorms which generates the topology τ od E.

Without loss of a generality we may assume that $q_{n+1} \geq q_n$, pointwise, for $n \in \mathbb{N}$.

Define

$$q_n^C(f) := \sup_{x \in \mathbb{R}} q_n(f(x)), \quad n \in \mathbb{N}.$$

Obviously (q_n^C) form a family of seminorms of $C(\mathbb{R}, E)$. Moreover, it is clear that $q_{n+1}^C \geq q_n^C$ for $n \in \mathbb{N}$.

Define the pseudo-norm

$$|f| = \sum_{n=1}^{\infty} \frac{1}{2^n} \frac{q_n^C(f)}{1 + q_n^C(f)} \quad \text{for} \quad f \in C(\mathbb{R}, E).$$

Obviously $C(\mathbb{R}, E)$ with the above defined pseudo-norm is a Fréchet space.

Now it is clear that $AP(E)$ is a linear subspace of $C(\mathbb{R}, E)$. In view of *Theorem 1.60 ii)* it is closed. This completes the proof. $\square$

Corollary 1.63. *If E is a Banach space, then the linear space of all almost periodic functions $\mathbb{R} \to E$ is a Banach space with the norm* sup.

We have also the following simple fact.

Proposition 1.64. *Let E be a Fréchet space over the field K ($K = \mathbb{R}$ or $\mathbb{C}$) and assume $f \in AP(E)$ and $\nu \in AP(K)$.*

Then $\nu f \in AP(E)$.

Proof: It is a simple consequence of the Bochner's criterion *Theorem 1.61.*

Definition 1.65. *A Fréchet space E is said to be perfect if every bounded function $f : \mathbb{R} \to E$ with an almost periodic derivative f' is necesssarily almost periodic.*

Does there exist a perfect Fréchet space which is not a Banach space?

The answer to this question is positive what we illustrate by the following.

Example 1.66. Denote s the linear space of all real sequences: $s = \{x = (x_n) : x_n \in \mathbb{N} \text{ for } n \in \mathbb{N}\}$.

For each $n \in \mathbb{N}$, define $p_n(x) = |x_n|$, $x \in s$. Obviously p_n is a seminorm defined on s.

Define $q_n := p_1 \vee p_2 \vee \ldots \vee p_n$ for $n \in \mathbb{N}$. We have $q_{n+1} \geq q_n$ for $n \in \mathbb{N}$.

The space s considered with the family of seminorms (q_n) is a Fréchet space.

Moreover, it can be proved (see [1] *Theorem 17.7*, p.210) that each closed and bounded subset of s is compact. Thus, in particular, s is not a Banach space.

Finally, in view of *Theorem 3.2.6* [80], s is perfect.

It is also possible to enlarge *Definition 1.58* to functions of two variables of the form $f(t, x)$ (see for instance [16]) as follows.

Definition 1.67. *A continuous function $f : \mathbb{R} \times E \to E$ is said to be almost periodic in t for each $x \in E$, if for each neighbourhood of the origin U, there exists a real number $l > 0$, such that every compact interval of the real line contains at least a point τ such that*

$$f(t + \tau, x) - f(t, x) \in U, \quad for \quad each \quad t \in \mathbb{R} \quad and \quad x \in E.$$

This definition is equivalent to the following, in view of the Bochner's Criterion.

Definition 1.68. *A continuous function $f(t,x) : \mathbb{R} \times E \to E$ is almost periodic in t for each $x \in E$ if for every sequence of real numbers (s_n'), there exists a subsequence (s_n) such that the sequence $(f(t+s_n, x))$ is uniformly convergent in $t \in \mathbb{R}$ and $x \in E$.*

We finally recall the useful result [16] *Lemma 3.8.*

Theorem 1.69. *Let $f : \mathbb{R} \times E \to E$ be almost periodic in t for each $t \in E$, and assume that f satisfies a Lipschitz condition in x uniformly in $t \in \mathbb{R}$, that is*

$$\rho(f(t,x), f(t,y)) \le L\rho(x,y),$$

for all $t \in \mathbb{R}$ and x, $y \in E$, where ρ is a metric on E. Let $\phi : \mathbb{R} \to E$ be almost periodic.

Then the function $F : \mathbb{R} \to E$ defined by $F(t) = f(t, \phi(t))$ is almost periodic.

1.8 Bibliographical Remarks and Open Problems

The study of almost automorphic functions with values in a Banach space was initiated by M. Zaki in his Ph. D. Thesis conducted under the supervision of S. Zaidman and published in the important paper [101]. After various subsequent contributions by S. Zaidman, B. Basit, G. M. N'Guérékata and many others, a systematic presentation has been made for the first time in [80].

To complete the study of vector-valued almost automorphic , it would be of great interest to establish the harmonic analysis of vector-valued almost automorphic functions. To this end we like to indicate the introduction of the notion of uniform spectrum by T. Diagana , M.V. Nguyen and G. M. NGuérékata in their recent and very important paper [32] that applies to almost autmorphic functions as well.

The Memoirs [88] from W. Shen and Y. Yi are to be cited among the most important contributions to the study of almost autmorphic functions and their applications to ordinary differential equations and dynamical systems.

Theorem 1.37 has a weak version. Indeed, the conclusion holds true even if $x(t)$ is weakly almost automorphic (see [79]).

It would be interesting to pursue the study in [80] *Section 2.2* in connection with the equation of vibrating string and Probability theory.

Section 1.7, including the notion of asymptotically almost periodic functions with values in a Fréchet space, is contained in a broader work by D. Bugajewski and G. M. N'Guérékata presented for the first time in [16]. This paper contains also many interesting applications to nonlinear differential and integro differential equations in Fréchet spaces.

The definition of asymptotically almost automorphic function used here might be generalized in taking for instance the corrective term as $h \in C(\mathbb{R}, X)$ and $\lim_{T \to \infty} \frac{1}{2T} \int_{-T}^{T} \|h(t)\| dt = 0$.

2

Almost Automorphic Evolution Equations

This chapter presents some of the most recent developments of the applications of almost automorphy to evolution equations. Some existence theorems are presented along with new methods to study almost automorphic solutions to linear and nonlinear evolution equations.

2.1 Linear Equations

2.1.1 The inhomogeneous equation $x' = Ax + f$

We consider in a Banach space $(X, \|.\|)$ the differential equation

$$x'(t) = Ax(t) + f(t), \quad t \in \mathbb{R}. \tag{2.1}$$

We will present various conditions for ensuring almost automorphy of the classical and/or mild solutions.

We start with the simplest case $A = \lambda \in \mathbb{C}$.

Theorem 2.1. *Let X be a uniformly convex (complex) Banach space. Suppose $f \in AA(X)$.*

Then every bounded solution of (2.1) is in $AA(X)$.

Proof: It is easy to check that equation (2.1) admits solutions of the form

$$x_1(t) = -\int_t^\infty e^{\lambda(t-r)} f(r)\, dr, \quad \text{if } \operatorname{Re}\lambda > 0$$

and

$$x_2(t) = \int_{-\infty}^t e^{\lambda(t-r)} f(r)\, dr, \quad \text{if } \operatorname{Re}\lambda < 0.$$

Let us prove that they are almost automorphic. We start with $x_1(t)$.

Let $s = t - r$; then we can write

$$x_1(t) = -\int_{-\infty}^0 e^{\lambda s} f(t - s)\, ds.$$

Let (s'_n) be an arbitrary sequence of real numbers. Since f is almost automorphic, there exists a subsequence $(s_n) \subseteq (s'_n)$ such that

$$\lim_{n\to\infty} f(t + s_n) = g(t)$$

and

$$\lim_{n\to\infty} g(t - s_n) = f(t)$$

pointwise on $\mathbb{R}$. So that, if we fix $t \in \mathbb{R}$, we can say that

$$\lim_{n\to\infty} f(t - s + s_n) = g(t - s)$$

for each $s \in \mathbb{R}$. We have

$$x_1(t + s_n) = -\int_{-\infty}^0 e^{\lambda s} f(t - s + s_n)\, ds.$$

Let us observe that

$$\|e^{\lambda s} f(t - s + s_n)\| \le e^{(\operatorname{Re}\lambda)s} \cdot \sup_{t\in\mathbb{R}} \|f(t)\|$$

and note that the right hand side of the above inequality is in $L^1(-\infty, 0)$ since for x, we consider $\operatorname{Re} \lambda > 0$. Now if we apply the Lebesgue's Dominated Convergence theorem (*Theorem 1.9*), since g is bounded and measurable over $\mathbb{R}$, we get

$$\lim_{n \to \infty} x_1(t + s_n) = -\int_{-\infty}^{0} e^{\lambda s} g(t - s) \, ds := y(t)$$

for each $t \in \mathbb{R}$. We can apply the same reasoning to obtain

$$\lim_{n \to \infty} y(t - s_n) = x_1(t),$$

for each t, which proves almost automorphy of $x_1(t)$.

The proof of almost automorphy of $x_2(t)$ is analogous.

Now if $\lambda = i\theta$ a pure imaginary number, then we get

$$x(t) = e^{it\theta} x(0) + \int_0^t e^{is\theta} f(s) ds, \quad t \in \mathbb{R}$$

Clearly $e^{is\theta}$ is $\mathbb{C}$- almost automorphic since it is almost periodic. Therefore $e^{is\theta} f(s)$ is almost automorphic too (*Proposition 1.35*).

Now $\int_0^t e^{is\theta} f(s) ds$ is bounded in X for $x(t)$ is bounded. And since X is uniformly convex, we deduce that it is almost automorphic. So $x \in AA(X)$ as the sum of two almost automorphic functions. The proof is now complete. □

Before we state our next result, let us recall the following definition:

Definition 2.2. *A linear operator $A : D(A) \subseteq X \to X$ where X is a (complex) Banach space is said to be of simplest type if $A \in L(X)$ the space of bounded linear operators in X, and*

$A = \sum_{k=1}^{n} \lambda_k P_k$, where the complex numbers λ_k are mutually distinct, and $(P_k)_{1 \leq k \leq n}$ forms a complex system $\sum_{k=1}^{n} P_k = I$ of mutually disjoint projections in X, that is $P_j P_k = \delta_{jk} P_k$, where δ_{jk} represents the Kronecker's symbol.

Now we have:

Theorem 2.3. *Assume $f \in AA(X)$ where X is a uniformly convex Banach space and A is of simplest type.*

Then every bounded solution of (2.1) is in $AA(X)$.

Proof. Take the projection P_k and apply it to equation (2.1). We get

$$
\begin{aligned}
P_k x'(t) &= \frac{d}{dt}(P_k x)(t) \\
&= P_k \Big(\sum_{j=1}^{n} \lambda_j P_j \Big) x(t) + P_k f(t) \\
&= \lambda_k (P_k x)(t) + (P_k f)(t).
\end{aligned}
$$

Clearly $P_k f \in AA(X)$ by *Theorem 2.3.6* [80], since $P_k \in L(X)$. Therefore by *Theorem 2.1*, $P_k x \in AA(X)$. We conclude that

$$
x(t) = \sum_{k=1}^{n} P_k x(t) \in AA(X)
$$

as a finite sum of almost automorphic functions. $\square$.

We now present the following reduction method:

Theorem 2.4. *Let A be a linear operator $\mathbb{C}^n \to \mathbb{C}^n$ and $f(t) : \mathbb{R} \to \mathbb{C}^n$ an almost automorphic function.*

Then every bounded solution $x(t)$ of (2.1) is almost automorphic.

Proof: Using *Proposition 1.3.7* in [80], we let B be an invertible linear operator $\mathbb{C}^n \to \mathbb{C}^n$ such that $B^{-1}AB$ is triangular with the representation

$$B^{-1}AB = \begin{pmatrix} \lambda_1 & c_{12} & \cdots & c_{1n} \\ 0 & \lambda_2 & \cdots & c_{2n} \\ \vdots & & & \\ 0 & 0 & \cdots & \lambda_n \end{pmatrix}$$

where $\lambda_1, \ldots, \lambda_n$ are the eigenvalues of A.

Let $x(t)$ be a bounded solution of (2.1) and put $y(t) = B^{-1}x(t)$. Then $y(t)$ is also bounded and it satisfies the equation

$$\begin{aligned} y'(t) &= B^{-1}x'(t) \\ &= B^{-1}Ax(t) + B^{-1}f(t) \\ &= B^{-1}ABy(t) + B^{-1}f(t). \end{aligned}$$

It is observed that $B^{-1}f(t) : \mathbb{R} \to \mathbb{C}^n$ is almost automorphic since B^{-1} is a bounded linear operator (*Theorem 2.3.6* [80]).

We can write the above equation as follows:

$$\begin{aligned} y_1'(t) &= \lambda_1 y_1(t) + c_{12}y_2(t) + \cdots + c_{1n}y_n(t) + g_1(t) \\ y_2'(t) &= \ldots\ldots\ldots \lambda_2 y_2(t) + \cdots + c_{2n}y_n(t) + g_2(t) \\ & \ldots\ldots\ldots\ldots\ldots\ldots\ldots\ldots\ldots \\ y_{n-1}' &= \ldots\ldots\ldots\ldots \lambda_{n-1}y_{n-1}(t) + c_{n-1\ n}y_n(t) + g_{n-1}(t) \\ y_n'(t) &= \ldots\ldots\ldots\ldots\ldots \lambda_n y_n(t) + g_n(t) \end{aligned}$$

where

$$y(t) = (y_1(t), \ldots, y_n(t)) \in \mathbb{C}^n$$

and

$$(g_1(t), \ldots, g_n(t)) = B^{-1}f(t), \quad t \in \mathbb{R}.$$

Now, $y_n(t)$ is an almost automorphic solution to the last equation (*Theorem 2.1*). We can say that $y_{n-1}(t)$ is also almost automorphic and proceed until $y_1(t)$, which proves that $y \in AA(X)$ and consequently $x = By \in AA(X)$ too. $\quad\square$

2.1.2 Method of Invariant Subspaces

In a Hilbert space $(H, \|\cdot\|)$, we consider the differential equation

$$x'(t) = (A + B)x(t), \quad t \in \mathbb{R}, \tag{2.2}$$

where both $A : D(A) \subset H \mapsto H$ and $B : D(B) \subset H \mapsto H$ are densely defined closed linear operators.

In *Section 4.4* [80], equation (2.2) is studied in the case where the operator B is bounded, using the so-called method of "decomposition of spaces".

Now, when B is unbounded, (that is we have un unbounded perturbation of the almost automorphic equation $x' = Ax$), this method is no longer efficient. We note that in general the algebraic sum $A+B$ is not always well-defined. Let's present some examples (see [26, 27]):

Example: Let $H = L^2(\mathbb{R})$ and consider the operators A and B defined by

$$Au = -u'', \quad D(A) = H^2(\mathbb{R})$$

and

$$Bu = Qu, \quad D(B) = \{u \in L^2(\mathbb{R}) : Qu \in L^2(\mathbb{R})\}.$$

Assume also that $Q : \mathbb{R} \to \mathbb{R}$ is a measurable function satisfying the assumption:

(H) $Q(x) \geq 0$, $Q \in L^1(\mathbb{R})$, and $Q \notin L^2_{loc}(\mathbb{R})$ (ie., $\int_a^b |f|^2 dx = \infty$ for every compact $[a, b] \subset \mathbb{R}$).

Then we have:

Proposition 2.5. *[26] Under assumption (H), $D(A) \cap D(B) = \{0\}$.*

Proof. Let $u \in D(A) \cap D(B)$ and suppose that $u \not\equiv 0$. u is continuous since it is in $H^2(\mathbb{R})$. Therefore there exist $I \subset \mathbb{R}$ an open subset and $\delta > 0$ such that $|u(x)| > \delta$ for all $x \in I$.

Now let $I' \subset I$ be a compact subset equipped with the induced topology on I.

Clearly: $Q_{|_{I'}} = \dfrac{(Q|u|)_{|_{I'}}}{|u|_{|_{I'}}} \in L^2(I')$, since $(Q|u|)_{|_{I'}} \in L^2(I')$ and $\dfrac{1}{|u|_{|_{I'}}} \in L^\infty(I')$.

As a consequence $Q \in L^2(I')$. This is impossible since $Q \notin L^2_{loc}(\mathbb{R})$.

Therefore $u \equiv 0$.

Example of Potential Q satisfying (H): Assume that the function f satisfies:

$f \geq 0$, $f \in L^2(\mathbb{R})$ with f^4 not integrable near $x = 0$.

Let $(\sigma_n)_{n=1}^\infty$ be an enumeration of the rational numbers, set

$$Q(x) = \sum_{k=1}^\infty k^{-2} f(x - \sigma_k),$$

satisfies the assumption (H).

Generalization. Let $H = L^2(\mathbb{R}^n)$ and let A, B be the operators given by

$$Au = -\Delta u, \quad D(A) = H^2(\mathbb{R}^n)$$

and

$$Bu = Qu, \quad D(B) = \{u \in L^2(\mathbb{R}^n) : Qu \in L^2(\mathbb{R}^n)\},$$

where $Q : \mathbb{R}^n \mapsto \mathbb{C}$ is a measurable function satisfying

(H)   Re $Q(x) \geq 0$, $Q \in L^1(\mathbb{R}^n)$, and $Q \notin L^2_{loc}(\mathbb{R}^n)$ (ie., $\int_\Omega |f|^2 dx = \infty$ for every compact $\Omega \subset \mathbb{R}^n$).

Then:

1. If $n \leq 3$, the previous proposition is still valid. One can prove it using a similar method as in the proof of *Proposition 2.5* .

2. If $n \geq 4$, the previous proposition is still valid (the Sobolev theorem implicitly used in the proof of *Proposition 2.5* cannot be used here anymore). This can be proved using the boundedness of the fractional operator $I_\alpha u := (-\Delta)^{\alpha/2} u$, when

a. $I_\alpha : L^2(\mathbb{R}^n) \mapsto BMO(\mathbb{R}^n)$

and

b. $I_\alpha : L^q(\mathbb{R}^n) \mapsto L^p(\mathbb{R}^n)$, where $1/q = 1/p - \alpha/n$ with $1/p > \alpha/n$.

Now, to overcome this difficulty, we will use the so-called "method of invariant spaces" introduced in [30].

But first, we will recall some preliminary facts:

Let $S \subset H$ be a closed subset and P_S, the orthogonal projection onto the subspace S. The operator is still a densely defined closed (possibly unbounded) linear operator in H.

Definition 2.6. *S is said to be an invariant subspace for A if we have the inclusion $A(D(A) \cap S) \subset S$.*

Example 2.7. Let us mention the following classical invariant subspaces for the closed unbounded linear operator A defined into the Hilbert space H.

1. $S = N(A) = \{x \in D(A) : Ax = 0\}$ is an invariant subspace for A.

2. If A is a self-adjoint linear operator, then any eigenspace $S_\lambda = N(\lambda I - A)$ is an invariant for A. In fact it can be easily shown that S_λ reduces A.

Theorem 2.8. *The equality $P_S A P_S = A P_S$ is a necessary and sufficient condition for a subspace S to be invariant for a linear operator A.*

Proof. Assume $P_S A P_S = A P_S$ and if $x \in D(A) \cap S$, then $x = P_S x \in D(A)$ and $Ax = A P_S x = P_S A P_S x \in S$.

Conversely, if S is invariant for A; let $x \in H$ be such that $P_S x \in D(A)$. Then $A P_S x \in S$ and then $P_S A P_S x = A P_S x$. Therefore $A P_S \subset P_S A P_S$. Since $D(A P_S) = D(P_S A P_S)$, it turns out that $A P_S = P_S A P_S$.

Definition 2.9. *A closed proper subspace S of the Hilbert space H is said to reduce an operator A if $P_S D(A) \subset D(A)$ and both S and $H \ominus S$, the orthogonal complement of S, are invariant for A.*

Using the above *Theorem 2.8*, the following key result can be proved.

Theorem 2.10. *A closed subspace S of H reduces an operator A if and only if $P_S A \subset A P_S$.*

Proof. See the proof in [59] *Theorem 4.11., p. 29.*

Remark 2.11. In fact the meaning of the inclusion $P_S A \subset A P_S$ is that: if $x \in D(A)$, then $P_S x \in D(A)$ and $P_S A x = A P_S x$.

Recall that their algebraic sum of A and B is defined by

$$D(A + B) = D(A) \cap D(B) \ \text{ and } \ (A + B)u = Au + Bu,$$

$\forall u \in D(A) \cap D(B)$.

We assume that $\overline{D(A) \cap D(B)} = H$ and the operators A and B are infinitesimal generators of C_0-groups of bounded linear operators $(T(t))_{t \in \mathbb{R}}$, $(R(t))_{t \in \mathbb{R}}$, respectively, such that

(i) $T(t)x : \ t \mapsto T(t)x$ is almost automorphic for each $x \in H$,
$R(t)y : \ t \mapsto R(t)y$ is almost automorphic for each $y \in H$, respectively;

(ii) there exists $S \subset H$, a closed subspace that reduces A and B. We denote by P_S, $Q_S = (I - P_S) = P_{H \ominus S}$, the orthogonal projections onto S and $H \ominus S$, respectively;

(iii) $R(A) \subseteq R(P_S) = N(Q_S)$;

(iv) $R(B) \subset R(Q_S) = N(P_S)$.

Remark 2.12. (1) Recall that if A, B generate C_0-groups, their sum $A + B$ need not be a C_0-group generator.

(2) The assumption (ii) above implies that both S and $H \ominus S$ are invariant for the algebraic sum (it is well-defined as stated above)$A + B$.

Theorem 2.13. *Under assumptions (i)-(ii)-(iii)-(iv), every solution to the differential equation (2.2) is almost automorphic.*

Proof. Let $x(s)$ be a solution to (2.2). Clearly $x(s) \in D(A) \cap D(B)$ (notice that the algebraic sum $A + B$ does exist since $A + B$ is assumed to be densely defined).

Now decompose $x(s)$ as follows

$$x(s) = x_1(s) + x_2(s), \tag{2.3}$$

where

$$x_1(s) = P_S x(s) \in R(P_S) = N(Q_S)$$

and

$$x_2(s) = Q_S x(s) \in N(P_S) = R(Q_S).$$

We have

$$
\begin{aligned}
\frac{d}{ds}(x_1(s)) &= P_S \frac{d}{ds} x(s) \\
&= P_S A x(s) + P_S B x(s) \\
&= A P_S x(s) + P_S B x(s) \quad \text{(according to}(ii)) \\
&= A x_1(s) + P_S B x(s) \\
&= A x_1(s) \quad \text{(according to}(iv)).
\end{aligned}
$$

From $\frac{d}{ds}(x_1(s)) = A x_1(s)$, it follows that

$$x_1(s) = T(s) x_1(0). \tag{2.4}$$

Now according to (i), the vector-valued function $s \mapsto T(s)x_1(0)$ is almost automorphic.

In the same way, since $H \ominus (H \ominus S) = S$. It follows that the closed subspace S reduces A and B if and only if $H \ominus S$ does. In other words, $H \ominus S$ reduces A and B. That is, a similar remark as *Remark 2.11* holds when S is replaced by $H \ominus S$. Thus, we have

$$
\begin{aligned}
\frac{d}{ds}(x_2(s)) &= Q_S \frac{d}{ds} x(s) \\
&= Q_S A x(s) + Q_S B x(s) \\
&= Q_S A x(s) + B Q_S x(s) \quad \text{(according to}(ii)) \\
&= Q_S A x(s) + B x_2(s) \\
&= B x_2(s) \quad \text{(according to}(iii)).
\end{aligned}
$$

From the equation $\frac{d}{ds}(x_2(s)) = B x_2(s)$, it follows that $s \mapsto R(s)x_2(0)$ is almost automorphic (according to (i)).

Therefore $x(s) = x_1(s) + x_2(s)$ is also almost automorphic as the sum of almost automorphic vector-valued functions.

Corollary 2.14. *Let $B : H \mapsto H$ be a bounded linear operator in the Hilbert space H. Under assumptions (i)-(ii)-(iii)-(iv), every solution to the equation (2.2) is almost automorphic.*

Proof. This an immediate consequence of the previous *Theorem 2.13* to the case where B is a bounded linear operator, it is straightforward.

2.1.3 Almost Automorphic Solutions to Some Second-Order Hyperbolic Equations

Consider now as in [28], the homogeneous second-order hyperbolic differential equation of the form

$$\frac{d^2}{ds^2}u(s) + 2B\,\frac{d}{ds}u(s) + A\,u(s) = 0, \tag{2.5}$$

and the associated nonhomogeneous differential equation

$$\frac{d^2}{ds^2}u(s) + 2B\,\frac{d}{ds}u(s) + A\,u(s) = f(s), \tag{2.6}$$

where A, B are densely defined closed linear operators acting in a Hilbert space H and $f : \mathbb{R} \mapsto H$ is an almost automorphic vector-valued function.

The method of invariant subspaces described in *Section 2.1.2* above can be used to deal with (2.5)-(2.6). For that, the idea is to reduce (2.5)-(2.6) into differential equations of first-order.

Indeed, setting $v(s) = \dfrac{d}{ds}u(s)$, the problem (2.5)-(2.6) can be rewritten in $H \times H$ of the form

$$\frac{d}{ds}\mathcal{U}(s) = (\mathcal{A} + \mathcal{B})\,\mathcal{U}(s), \tag{2.7}$$

and

$$\frac{d}{ds}\mathcal{U}(s) = (\mathcal{A} + \mathcal{B})\,\mathcal{U}(s) + F(s), \tag{2.8}$$

where $\mathcal{U}(s) = (u(s), v(s))$, $F(s) = (0, f(s))$ and $\mathcal{A}, \mathcal{B}$ are the operator matrices of the form

$$\mathcal{A} = \begin{pmatrix} O & I \\ -A & O \end{pmatrix} \quad \text{and} \quad \mathcal{B} = \begin{pmatrix} O & O \\ O & -2B \end{pmatrix},$$

on $H \times H$ with $D(\mathcal{A}) = D(A) \times H$, $D(\mathcal{B}) = H \times D(B)$, and O, I denote the zero and identity operators on H, respectively.

Since (2.5)-(2.6) is equivalent to (2.7)-(2.8), instead of studying (2.5)-(2.6), we will focus on the characterization of almost automorphic solutions to (2.7)-(2.8).

In this book, we will treat only the homogeneous case (2.7). We refer the reader to [28] for the nonhomogeneous case (2.8).

As in the previous section, we will make the following assumptions:

The operators $\mathcal{A}$ and $\mathcal{B}$ are infinitesimal generators of C_0-groups of bounded operators $(\mathcal{T}(t))_{s \in \mathbb{R}}$, $(\mathcal{R}(t))_{s \in \mathbb{R}}$, respectively, such that

(i) $\mathcal{T}(s)\mathcal{U} : s \mapsto \mathcal{T}(s)\mathcal{U}$ is almost automorphic for each $\mathcal{U} \in H \times H$, $\mathcal{R}(s)\mathcal{V} : s \mapsto \mathcal{R}(s)\mathcal{V}$ is almost automorphic for each $\mathcal{V} \in H \times H$, respectively;

(ii) There exists $S \subset H \times H$, a closed subspace that reduces both $\mathcal{A}$ and $\mathcal{B}$.

We denote by $P_S, Q_S = (I \times I - P_S) = P_{[H \times H] \ominus S}$, the orthogonal projections onto S and $[H \times H] \ominus S$, respectively;

(iii) $R(\mathcal{A}) \subset R(P_S) = N(Q_S)$;

(iv) $R(\mathcal{B}) \subset R(Q_S) = N(P_S)$.

We have

Theorem 2.15. *Under assumptions (i)-(ii)-(iii)-(iv), every solution to the differential equation (2.7) is almost automorphic.*

Proof. Let $X(s)$ be a solution to (2.7).

Clearly $X(s) \in D(\mathcal{A}) \cap D(\mathcal{B}) = D(A) \times D(B)$ (observe that the algebraic sum $\mathcal{A} + \mathcal{B}$ exists since $\overline{D(\mathcal{A} + \mathcal{B})} = H \times H$).

Now decompose $X(s)$ as follows

$$X(s) = P_S X(s) + (I \times I - P_S)X(s),$$

where $P_S X(s) \in R(P_S) = N(Q_S)$, and $Q_S X(s) \in N(P_S) = R(Q_S)$.

We have

$$\begin{aligned}
\frac{d}{ds}(P_S X(s)) &= P_S \frac{d}{ds} X(s) \\
&= P_S \mathcal{A} X(s) + P_S \mathcal{B} X(s) \\
&= \mathcal{A} P_S X(s) + P_S \mathcal{B} X(s) \quad \text{(by} \quad (ii)) \\
&= \mathcal{A} P_S X(s) \quad \text{(by} \quad (iv))
\end{aligned}$$

From $\frac{d}{ds}(P_S X(s)) = \mathcal{A} P_S X(s)$, it follows that

$$P_S X(s) = \mathcal{T}(t) P_S X(0).$$

Now according to (i), the vector-valued function $s \mapsto P_S X(s) = \mathcal{T}(t)P_S X(0)$ is almost automorphic.

In the same way, since we have $[H \times H] \ominus ([H \times H] \ominus S) = S$, then it follows that the closed subspace S reduces $\mathcal{A}$ and $\mathcal{B}$ if and only if $[H \times H] \ominus S$ does.

Hence

$$\begin{aligned}
\frac{d}{ds}(Q_S X(s)) &= Q_S \frac{d}{ds} X(s) \\
&= Q_S \mathcal{A} X(s) + Q_S \mathcal{B} X(s) \\
&= Q_S \mathcal{A} X(s) + \mathcal{B} Q_S X(s) \quad \text{(by} \quad (ii)) \\
&= \mathcal{B} Q_S X(s) \quad \text{(by} \quad (iii))
\end{aligned}$$

From the equation $\dfrac{d}{dt}(Q_S X(s)) = \mathcal{B} Q_S X(s)$, it follows that $s \mapsto Q_S X(s) = \mathcal{R}(s) Q_S X(0)$ is almost automorphic (according to (i)).

Therefore $X(s) = P_S X(s) + Q_S X(s)$ is also almost automorphic as the sum of almost automorphic vector-valued functions.

Corollary 2.16. *Let $B : H \mapsto H$ be a bounded linear operator in the Hilbert space H. Under assumptions (i)-(ii)-(iii)-(iv). Then every solution to the equation (2.7) is almost automorphic.*

Proof. This an immediate consequence of the *Theorem 2.15* to the case where $\mathcal{B}$ is a bounded linear operator, it is straightforward.

2.2 Nonlinear Equations

2.2.1 Existence of Almost Automorphic Mild Solutions-Case I

We first begin with the following semilinear evolution equation in a Banach space $(\mathbb{X}, \| \cdot \|)$:

$$x'(t) = Ax(t) + f(t, x(t)), \quad t \in \mathbb{R} \qquad (2.9)$$

where A is the infinitesimal generator of an exponentially stable C_0-semigroup $(T(t))_{t \geq 0}$; that is, there exist $K > 0$, $\omega < 0$ such that

$$\|T(t)\| \leq K e^{\omega t}, \quad \text{for all } t \geq 0. \qquad (2.10)$$

We assume that $f : \mathbb{R} \times X \mapsto X$ satisfies a Lipschitz condition in x uniformly in t, that is, there exists $L > 0$ such that

$$\|f(t,x) - f(t,y)\| \le L\|x - y\|$$

for every $t \in \mathbb{R}$, $x, y \in X$.

We wish to establish existence and uniqueness of almost automorphic mild solutions to the equation (2.9). We first prove the existence of almost automorphic mild solution of the differential equation

$$x'(t) = Ax(t) + f(t), \quad t \in \mathbb{R}, \tag{2.11}$$

where $f \in AA(X)$. In fact we have.

Theorem 2.17. *Let $f \in AA(X)$ and A be the generator of an exponentially stable C_0-semigroup as above .*

Then equation (2.11) has a unique almost automorphic mild solution.

Proof. We first prove existence of an almost automorhic solution.

Let

$$x(t) = T(t-a)x(a) + \int_a^t T(t-s)f(s)ds, \text{ for all } a \in \mathbb{R}, \ t \ge a,$$

be a mild solution. It remains to prove that $x \in AA(X)$.

First, we consider $u(t) = \int_{-\infty}^t T(t-s)f(s)ds$, defined as

$$\lim_{r \downarrow -\infty} \int_r^t T(t-s)f(s)ds.$$

Clearly for each $r < t$, the integral $\int_r^t T(t-s)f(s)ds$ exists. Moreover

$$\left\| \int_r^t T(t-s)f(s)ds \right\| \leq \frac{K}{|\omega|} \|f\|_\infty, \quad for \ all \ r < t.$$

which shows $\int_{-\infty}^t T(t-s)f(s)ds$ is absolutely convergent.

Now let (s_n') be an arbitrary sequence of real numbers. Since $f \in AA(X)$, there exists a subsequence (s_n) of (s_n') such that

$$g(t) := \lim_{n \to +\infty} f(t + s_n)$$

is well-defined for each $t \in \mathbb{R}$ and

$$f(t) = \lim_{n \to +\infty} g(t - s_n)$$

for each $t \in \mathbb{R}$

Now consider

$$
\begin{aligned}
u(t + s_n) &= \int_{-\infty}^{t+s_n} T(t + s_n - s)f(s)ds \\
&= \int_{-\infty}^t T(t - \sigma)f(\sigma + s_n)d\sigma \\
&= \int_{-\infty}^t T(t - \sigma)f_n(\sigma)d\sigma,
\end{aligned}
$$

where $f_n(\sigma) = f(\sigma + s_n), \quad n = 1, 2, ...$

We also have

$$\|u(t + s_n)\| \leq \frac{K}{|\omega|} \|f\|_\infty, \quad for \ all \ n = 1, 2, ...$$

and by continuity of the semigroup, $T(t-\sigma)f_n(\sigma) \mapsto T(t-\sigma)g(\sigma)$, as $n \to \infty$ for each $\sigma \in \mathbb{R}$ fixed and any $t \geq \sigma$.

If we let $v(t) = \int_{-\infty}^t T(t - s)g(s)ds$, we observe that the integral is absolutely convergent for each t. So, by the Lebesgue's Dominated Convergent Theorem (*Theorem 1.9*),

$$u(t + s_n) \rightarrow v(t), \quad \text{as} \ \ n \rightarrow \infty$$

for each $t \in \mathbb{R}$.

We can show in a similar way that

$$v(t - s_n) \rightarrow u(t) \ \ \text{as} \ \ n \rightarrow \infty$$

for each $t \in \mathbb{R}$. This shows that $u \in AA(X)$.

Now let $u(a) = \int_{-\infty}^{a} T(a - s)f(s)ds$.

So $T(t - a)u(a) = \int_{-\infty}^{a} T(t - s)f(s)ds$.

If $t \geq a$, then

$$\int_{a}^{t} T(t - s)f(s)ds = \int_{-\infty}^{t} T(t - s)f(s)ds - \int_{-\infty}^{a} T(t - s)f(s)ds$$
$$= u(t) - T(t - a)u(a).$$

so that, $u(t) = T(t - a)u(a) + \int_{a}^{t} T(t - s)f(s)ds$. If we fix $x(a) = u(a)$, then $x(t) = u(t)$, that is $x \in AA(X)$.

We finally prove the uniqueness of the almost automorphic solution.

Assume x and y are two such solutions and we let $z = x - y$. Then $z \in AA(X)$ and satisfies the equation

$$z'(t) = Az(t), \ \ t \in \mathbb{R}.$$

Note that z is bounded and satisfies also the equation

$$z(t) = T(t - s)z(s) \ \ \text{for all} \ \ s \in \mathbb{R}, \ and \ t \geq s$$

We also have the inequality

$$\|z(t)\| \leq Ke^{\omega(t-s)}.$$

Take a sequence of real numbers (s_n) such that $s_n \to -\infty$. For any fixed $t \in \mathbb{R}$, we then can find a subsequence (s_{n_k}) of (s_n) with $s_{n_k} < t$ for all $k = 1, 2, \ldots$. Using the fact that $\omega < 0$, we obtain $z = 0$.

This shows uniqueness of the solution and ends the proof.   □

We now state and prove:

Theorem 2.18. *Assume that $f : \mathbb{R} \times X \mapsto X$ satisfies a Lipschitz condition in x uniformly in t, that is,*

$$\|f(t,x) - f(t,y)\| \leq L\|x - y\|, \quad \text{for all} \ \ x, y \in X,$$

where $L < \dfrac{|\omega|}{K}$. Let also $f(t,x)$ be almost automorphic in t for each $x \in X$.

Then equation (2.9) has a unique almost automorphic mild solution.

Proof. Let x be a mild solution. It is continuous and satisfies the integral equation

$$x(t) = T(t - a)x(a) + \int_a^t T(t - s)f(s, x(s))ds, \quad \forall a \in \mathbb{R}, \ \forall \, t \geq a.$$

Consider now $\int_a^t T(t - s)f(s, x(s))ds$ and the nonlinear operator $G : AA(X) \mapsto AA(X)$ given by

$$(G\phi)(t) := \int_{-\infty}^t T(t - s)f(s, \phi(s))ds.$$

In view of *Theorem 2.2.6* in [80], if $\phi \in AA(X)$, then $f(s, \phi(s))$ is almost automorphic, thus $G\phi \in AA(X)$, so that G is well-defined.

Now for $\phi_1, \, \phi_2 \in AA(X)$, we have:

$$\|G\phi_1 - G\phi_2\|_\infty = \sup_{t\in\mathbb{R}} \left\| \int_{-\infty}^t T(t-s)\{f(s,\phi_1(s)) - f(s,\phi_2(s))\}ds \right\|$$

$$\leq \sup_{t\in\mathbb{R}} \int_{-\infty}^t \|T(t-s)\|_{B(X)} L\|\phi_1(s) - \phi_2(s)\|ds$$

$$\leq L\|\phi_1 - \phi_2\|_\infty \cdot \sup_{t\in\mathbb{R}} \int_{-\infty}^t \|T(t-s)\|_{B(X)}ds$$

$$\leq L\|\phi_1 - \phi_2\|_\infty \cdot \sup_{t\in\mathbb{R}} \int_{-\infty}^t K e^{\omega(t-s)}ds$$

$$= \frac{LK}{|\omega|}\|\phi_1 - \phi_2\|_\infty.$$

So

$$\|G\phi_1 - G\phi_2\|_\infty \leq \frac{LK}{|\omega|}\|\phi_1 - \phi_2\|_\infty,$$

which proves that G is continuous. And since $\dfrac{LK}{|\omega|} < 1$, then G is a contraction. So there exists a unique $u \in AA(X)$, such that $Gu = u$, that is $u(t) = \int_{-\infty}^t T(t-s)f(s,u(s))ds$.

If we let $u(a) = \int_{-\infty}^a T(a-s)f(s,u(s))ds$, then

$$T(t-a)u(a) = \int_{-\infty}^a T(t-s)f(s,u(s))ds.$$

But for $t \geq a$,

$$\int_a^t T(t-s)f(s,u(s))ds = \int_{-\infty}^t T(t-s)f(s,u(s))ds$$

$$= \int_{-\infty}^a T(t-s)f(s,u(s))ds$$

$$= u(t) - T(t-a)u(a).$$

So $u(t) = T(t-a)u(a) + \int_{-\infty}^t T(t-s)f(s,u(s))ds$ is a mild solution of equation (2.9) and $u \in AA(X)$. The proof is now complete. $\square$

2.2.2 Existence of Almost Automorphic Mild Solutions-Case II

We still consider equation (2.9) where A is described as above, but this time $f(t,x)$ does not satisfy necessarily a Lipschitz condition. In fact we wish to consider another class of functions $f(t,x)$ as described below.

First, let $(\mathbb{Y}, |\cdot|)$ denote a Banach space algebraically contained in $(\mathbb{X}, \|\cdot\|)$ such that the canonical injection $\mathbb{Y} \to \mathbb{X}$ is compact. An example of such a space $\mathbb{Y}$ is an abstract Sobolev space that we construct as follows:

Let A be as above. Then clearly $0 \in \rho(A)$, so that the fractional powers $(-A)^\alpha$, $0 < \alpha < 1$, are well defined. Also, since $0 \in \rho(A)$, the norm

$$|f| = \|(-A)^\alpha f\| \tag{2.12}$$

is equivalent to the graph norm

$$\|f\|_\alpha = \|(-A)^\alpha f\| + \|f\|.$$

Now we take $\mathbb{X} = L^p(\Omega)$, where $1 < p < \infty$ and $\Omega \subset \mathbb{R}^n$ is a smooth bounded domain in $\mathbb{R}^n$. Let A be a linear uniformly elliptic operator (with suitable boundary conditions), of order $2m$. Then let $\mathbb{Y}$ be the domain of $(-A)^\alpha$ with norm (2.12); we have

$$W_0^{2m\alpha,p}(\Omega) \subset \mathbb{Y} \subset W^{2m\alpha,p}(\Omega)$$

and the norm $|\cdot|$ in $\mathbb{Y}$ is equivalent to the usual norm in $W^{2m\alpha,p}(\Omega)$. Also, the injection $\mathbb{Y} \to \mathbb{X}$ is compact in this case, by Sobolev embedding.

Now let $\mathbb{Y} = D((-A)^\alpha)$, the domain of $(-A)^\alpha$, with norm

$$|y| = \|(-A)^\alpha y\|, \quad y \in D((-A)^\alpha),$$

where $0 < \alpha < 1$ is fixed. We get

$$|T(t)y| = \|T(t)(-A)^\alpha y\| \le Ke^{-\omega t}\|(-A)^\alpha y\|,$$

and since

$$Ke^{-\omega t}\|(-A)^\alpha y\| = Ke^{-\omega t}|y|,$$

we obtain

$$|T(t)y| \le Ke^{-\omega t}|y| \qquad (2.13)$$

for each $y \in \mathbb{Y}$ and every $t \ge 0$, by (2.10).

We also make the following assumptions:

$$F(t,x) = P(t)Q(x), \quad for\ all \quad t \in \mathbb{R}, x \in \mathbb{X}, \qquad (2.14)$$

where $P(t) \in AA(\mathbb{Z})$ for each $t \in \mathbb{R}$ with $\mathbb{Z} = B(\mathbb{X}, \mathbb{Y})$; P is continuous from $\mathbb{R}$ to $AA(\mathbb{Z})$, and $Q : BC(\mathbb{R}, \mathbb{X}) \to BC(\mathbb{R}, \mathbb{X})$ is continuous and satisfies the estimate

$$\|Q\varphi\|_\infty \le \mathcal{M}(\|\varphi\|_\infty), \qquad (2.15)$$

where $\|f\|_\infty := \sup_{t \in \mathbb{R}} \|f(t)\|$ and $\mathcal{M} \in C(\mathbb{R}^+, \mathbb{R}^+)$ satisfies

$$\lim_{r \to \infty} \frac{\mathcal{M}(r)}{r} = 0. \qquad (2.16)$$

Note that $\mathcal{M}$ can be unbounded but must grow slower than a linear function. Let

$$[P] := \sup_{t \in \mathbb{R}} \|P(t)\|_{\mathbb{Z}} < \infty. \tag{2.17}$$

Define $G : BC(\mathbb{R}, \mathbb{X}) \to BC(\mathbb{R}, \mathbb{Y})$ by

$$(G\varphi)(t) = \int_{-\infty}^{t} T(t-s)F(s, \varphi(s))ds. \tag{2.18}$$

For $\varphi \in BC(\mathbb{R}, \mathbb{X})$, this integral exists. Indeed, we have

$$|(G\varphi)(t)| \le \int_{-\infty}^{t} |T(t-s)||P(t)Q(\varphi(s))|ds$$

$$\le \int_{-\infty}^{t} Ke^{-\omega(t-s)}[P]\mathcal{M}(\|\varphi\|_\infty)ds$$

using (2.13), (2.14) and (2.17). Consequently

$$|G\varphi|_\infty = \sup_{t \in \mathbb{R}} |(G\varphi)(t)|$$

$$\le K\omega^{-1}[P]\mathcal{M}(\|\varphi\|_\infty). \tag{2.19}$$

Continuity of G is straightforward in virtue of continuity of both P and Q. Thus we have

$$G(BC(\mathbb{R}, \mathbb{X})) \subset BC(\mathbb{R}, \mathbb{Y}).$$

Finally, for $0 < \delta \le 1$, let

$$BC^\delta(\mathbb{R}, \mathbb{Y}) \equiv \{f \in BC(\mathbb{R}, \mathbb{Y}) : |f|_{\delta, \mathbb{Y}} < \infty\},$$

where

$$|f|_{\delta, \mathbb{Y}} \equiv \sup_{t \in \mathbb{R}} |f(t)| + \delta \sup_{t, s \in \mathbb{R}, t \ne s} \frac{|f(t) - f(s)|}{|t - s|^\delta}.$$

With the norm $|\cdot|_{\delta, \mathbb{Y}}$, $BC^\delta(\mathbb{R}, \mathbb{Y})$ turns out to be a Banach space of all bounded Hölder continuous $\mathbb{Y}$-valued functions on $\mathbb{R}$ of Hölder exponent δ.

Proposition 2.19. *The function G defined above maps bounded sets of $BC(\mathbb{R}, \mathbb{X})$ into bounded sets of $BC^\delta(\mathbb{R}, \mathbb{Y})$ for any $\delta > 0$ satisfying $\delta < \alpha$, where $0 < \alpha < 1$ is the exponent defining $\mathbb{Y} = D(-A)^{-\alpha}$.*

Proof. The proof is basically a modification of the above remarks. Let $0 < \beta < \alpha$. Then

$$|(G\varphi)(t)| \leq |\int_{-\infty}^{t} T(t-s)(-A)^\beta(-A)^{-\beta}F(s, \varphi(s))ds|$$

$$\leq \int_{-\infty}^{t} |T(t-s)(-A)^\beta||(-A)^{-\beta}P(s)||Q(\varphi(s))|ds \quad (2.20)$$

Now, by *Proposition 1.19*, there exists a constant K_1 such that

$$\|T(r)(-A)^\beta\| \leq \frac{K_1 e^{-\omega r}}{r^\beta}$$

for all $r > 0$. Thus we obtain, as previously,

$$|T(r)(-A)^\beta| \leq K_1 e^{-\omega r} r^{-\beta}, \quad r > 0. \quad (2.21)$$

Next, we observe that the function $s \mapsto (-A)^{-\beta}P(s)$ is a uniformly bounded function $\mathbb{R} \to B(\mathbb{X}, D((-A)^{\alpha-\beta}))$. Indeed, it is the composition of $P(\cdot) : \mathbb{R} \to B(\mathbb{X}, D((-A)^\alpha))$ which is bounded by $[P]$, with $(-A)^{-\beta}$, an isometry from $D((-A)^\alpha)$ onto $D((-A)^{\alpha-\beta})$. Thus

$$\sup_{t\in\mathbb{R}} \|P(t)\|_{B(\mathbb{X}, D((-A)^{\alpha-\beta}))} \leq [P].$$

Now combining the estimates in (2.20) and (2.21), we deduce

$$|(G\varphi)(t)| \leq \int_{-\infty}^{t} K_1 e^{-\omega(t-s)}(t-s)^{-\beta}[P]\mathcal{M}(\|P\|_\infty)ds.$$

Letting $r = t - s$ in the integral gives

$$|G\varphi(t)| \leq \int_0^\infty K_1 e^{-\omega r} r^{-\beta} [P] \mathcal{M}(\|\varphi\|) dr,$$

that is,

$$|(G\varphi)(t)| \leq C_1(\beta) \mathcal{M}(\|\varphi\|_\infty), \tag{2.22}$$

where $C_1(\beta)$ depends on β, K_1, ω and $[P]$.

Next, for $t_2 > t_1$, we have

$$|(G\varphi)(t_2) - (G\varphi)(t_1)| \leq$$

$$|(\int_{-\infty}^{t_2} - \int_{-\infty}^{t_1}) T(t_2 - s)(-A)^\beta (-A)^{-\beta} P(s) Q(\varphi(s)) ds|$$

$$+|\int_{-\infty}^{t_1} (T(t_2 - s) - T(t_1 - s))(-A)^\beta (-A)^{-\beta} P(s) Q(\varphi(s)) ds|$$

$$\leq \int_{t_1}^{t_2} |T(t_2 - s)(-A)^\beta (-A)^{-\beta} \times P(s) Q(\varphi(s))|$$

$$+\int_{-\infty}^{t_1} |(T(t_2 - t_1) - I) \times T(t_1 - s)(-A)^\beta (-A)^{-\beta} P(s) Q(\varphi(s))| ds$$

$$= J_1 + J_2.$$

By the same argument leading to (2.22) we get

$$J_1 \leq \int_0^{t_2 - t_1} K_1 e^{-\omega r} r^{-\beta} [P] \mathcal{M}(\|\varphi\|_\infty) dr$$
$$\leq C_2(\beta) \mathcal{M}(\|\varphi\|_\infty)(t_2 - t_1)^{1-\beta}.$$

Also, we have

$$J_2 \le \int_{\infty}^{t_1} |(T(t_2 - t_1) - I)(-A)^{-\gamma}(T(t_1 - s)(-A)^{(\beta-\gamma)}(-A)^{-\beta}$$
$$\times P(s)Q(\varphi(s))|ds$$
$$\le \int_{\infty}^{t_1} |(T(t_2 - t_1) - I)(-A)^{-\gamma}| \cdot |(T(t_1 - s)$$
$$\times (-A)^{(\beta-\gamma)}(-A)^{-\beta}P(s)Q(\varphi(s))|ds$$
$$\le |(T(t_2 - t_1) - I)(-A)^{-\gamma}|$$
$$\times \int^{t_1} |T(t_1 - s)(-A)^{(\beta-\gamma)}(-A)^{-\beta}P(s)Q(\varphi(s))|ds$$
$$\le |(T(t_2 - t_1) - I)(-A)^{-\gamma}|C_3(\beta,\gamma)\mathcal{M}(\|\varphi\|_\infty)$$

provided $0 < \gamma < \beta$.

Next recall that $(T(r) - I)g = \int_0^r T(s)Agds$ for $g \in D(A)$, by the fundamental theorem of calculus. Thus, for $f \in \mathbb{Y}$,

$$|(T(r) - I)(-A)^{-\gamma}f| = \| \int_0^r T(s)(-A)^{1-\gamma-\alpha}(-A)^\alpha f ds\|$$
$$\le \|(-A)^\alpha f\| \int_0^r M_1 e^{-\varepsilon s} s^{1-\gamma-\alpha} ds$$
$$= C_4(\gamma, \omega, K_1) r^{2-\gamma-\alpha} |f|,$$

since $1 - \gamma - \alpha > -1$, because $0 < \gamma < \beta < \alpha < 1$.

In other words, we have

$$|(T(r) - I)(-A)^{-\gamma}| \le C_4 r^{2-\gamma-\alpha};$$

consequently,

$$J_2 \le C_4(t_2 - t_1)^{2-\gamma-\alpha} C_3 \mathcal{M}(\|\varphi\|_\infty).$$

For $\delta = min(2 - \gamma - \alpha, 1 - \beta) > 0$, it follows that

$$|(G\varphi)(t_2) - (G\varphi(t_1)| \le C_5 |t_2 - t_1|^\delta \mathcal{M}(|\varphi\|_\infty), \qquad (2.23)$$

where C_5 depends on $\omega, K_1, [P], \alpha, \beta, \gamma$ and $\mathbb{Y}$, that is, on parameters of the problem.

It follows that, for $\varphi \in BC(\mathbb{R}, \mathbb{X})$ with $\|\varphi(t)\| \leq R$ for all $t \in \mathbb{R}$, then $G\varphi \in BC^\delta(\mathbb{R}, \mathbb{Y})$ with $\|G\varphi(t)\| \leq R_1$ for all $t \in \mathbb{R}$ and some R_1 that depends on R.

This completes the proof. $\square$

Proposition 2.20. *The function G maps bounded sets of $AA(\mathbb{X})$ into bounded sets of $BC^\delta(\mathbb{R}, \mathbb{Y}) \cap AA(\mathbb{X})$ for $0 < \delta < \alpha$.*

Proof. We just need to check that

$$G(AA(\mathbb{X})) \subset AA(\mathbb{X}).$$

To this end, let $\varphi \in AA(\mathbb{X})$. Then given a sequence $(s'_n) \subset \mathbb{R}$, there exists a subsequence $(s_n) \subset (s'_n)$ such that

$$\psi(t) = \lim_{n \to \infty} \varphi(t + s_n),$$

is well defined for each $t \in \mathbb{R}$ and

$$\lim_{n \to \infty} \psi(t - s_n) = \varphi(t)$$

for each $t \in \mathbb{R}$. Since $\psi \in BC(\mathbb{R}, \mathbb{X})$, then

$$(G\varphi)(t + s_n) = \int_{-\infty}^{t+s_n} T(t + s_n - s)P(s)Q(\varphi(s))ds.$$

Let $\sigma = s - s_n$. Then

$$(G\varphi)(t + s_n) = \int_{-\infty}^{t} T(t - \sigma)P(\sigma + s_n)Q(\varphi(\sigma + s_n))d\sigma$$

$$= \int_{-\infty}^{t} T(t - \sigma)P_n(\sigma)Q_n(\sigma)d\sigma,$$

where $P_n(\sigma) = P(\sigma + s_n), Q_n(\sigma) = Q((\varphi(\sigma + s_n))), \ n = 1, 2, \cdots,$ $\sigma \in \mathbb{R}.$

Since $P \in AA(\mathbb{Z})$, there exists a subsequence of (s_n), which we still denote by (s_n), such that

$$\hat{P}(\sigma) = \lim_{n \to \infty} P_n(\sigma)$$

exists for each $\sigma \in \mathbb{R}$ and

$$\lim_{n \to \infty} \hat{P}(\sigma - s_n) = P(\sigma)$$

for each $\sigma \in \mathbb{R}.$

Clearly we also have, by passing to a subsequence if necessary,

$$\lim_{n \to \infty} \varphi(t + s_n) = \psi(t)$$

and

$$\lim_{n \to \infty} \psi(t - s_n) = \varphi(t),$$

for each $t \in \mathbb{R}$. By the Bochner's integral version of the Lebesgue's Dominated Convergence theorem, we get

$$(G\varphi)(t + s_n) = \int_{-\infty}^{t} T(t - \sigma) P_n(\sigma) Q_n(\sigma) d\sigma$$

$$\longrightarrow \int_{-\infty}^{t} T(t - \sigma) \hat{P}(\sigma) Q(\varphi(\sigma)) d\sigma = \chi(t)$$

for each $t \in \mathbb{R}$, and

$$\chi(t - s_n) = \int_{-\infty}^{t - s_n} T(t - s_n - \sigma) \hat{P}(\sigma) Q(\psi(\sigma)) d\sigma$$

$$= \int_{-\infty}^{t} T(t - r) \hat{P}(r - s_n) Q(\psi(r - s_n)) dr$$

by letting $r = \sigma + s_n$, thus we obtain

$$\chi(t - s_n) \longrightarrow \int_{-\infty}^{t} T(t-r)P(r)Q(\varphi(r))dr = (G\varphi)(t),$$

again by the Lebesgue's Dominated Convergence theorem. This shows that

$$G(AA(\mathbb{X})) \subset AA(\mathbb{X})$$

and the proof is now complete.

Proposition 2.21. *The canonical injection* $id : BC^{\delta}(\mathbb{R}, \mathbb{Y}) \rightarrow BC(\mathbb{R}, \mathbb{X})$ *is compact, which implies that*

$$id : BC^{\delta}(\mathbb{R}, \mathbb{Y}) \cap AA(\mathbb{X}) \rightarrow AA(\mathbb{X})$$

is compact too.

Proof. We wish to show that id maps bounded sets of $BC^{\delta}(\mathbb{R}, \mathbb{Y})$ into relatively compact sets of $BC(\mathbb{R}, \mathbb{X})$.

To this end, let (φ_ν) be a bounded sequence in $BC^{\delta}(\mathbb{R}, \mathbb{Y})$.

Let $\mathbb{Q} = \{r_n\}$ be the set of all rational numbers. Then $(\varphi_\nu(r_n))$ is a bounded sequence in $\mathbb{Y}$, for each n.

By the well known Cantor diagonalization process, there exists a subsequence (φ_{ν_k}) such that

$$\varphi_{\nu_k}(r_n) \rightarrow \varphi(r_n),$$

as $k \rightarrow \infty$ in $\mathbb{X}$, for each n, and some $\varphi : \mathbb{Q} \rightarrow \mathbb{X}$.

But the sequence (φ_n) is an equicontinuous family of functions in $BUC(\mathbb{R}, \mathbb{Y}) \subset BUC(\mathbb{R}, \mathbb{X})$, because of the uniform Hölder condition.

Thus, as in the proof of the Arzela-Ascoli theorem, there is a further subsequence (which we still denote by (φ_{ν_k})) satisfying

$$\varphi_{\nu_k}(t) \to \varphi(t), \text{ as } k \to \infty \qquad (2.24)$$

in $\mathbb{X}$, for all $t \in \mathbb{R}$.

In addition the convergence is uniform in $t \in \mathbb{R}$.

Note that $BUC(\mathbb{R}, \mathbb{X})$ can be identified with $C(\Gamma, \mathbb{X})$ for a suitable Hausdorff compactification Γ of $\mathbb{R}$ (see for instance [33]).

Thus the convergence $\varphi_{\nu_k} \to \varphi$ holds in $BUC(\mathbb{R}, \mathbb{X}) \subset BC(\mathbb{R}, \mathbb{X})$.

This completes the proof.   □

Proposition 2.22. *The function G has a fixed point in $AA(\mathbb{X})$.*

Proof. Let us recall that the estimates (2.22)-(2.23),

$$|G\varphi|_\infty \leq C_1(\beta)\mathcal{M}(\|\varphi\|_\infty)$$

and

$$|(G\varphi)(t_2) - (G\varphi(t_1)|_\leq C_5|t_2 - t_1|^\delta \mathcal{M}(\|\varphi\|_\infty),$$

hold for all $\varphi \in BC(\mathbb{R}, \mathbb{Y})$ and all $t_1, t_2 \in \mathbb{R}$ with t_2 not equal to t_1.

It follows that there exists a constant $C_6 = C_6(\omega, K, K_1, \alpha, \beta, \gamma)$ such that

$$\varphi \in BC(\mathbb{R}, \mathbb{X}) \qquad \text{and} \qquad \|\varphi\|_\infty < R \text{ implies}$$
$$G\varphi \in BC^\delta(\mathbb{R}, \mathbb{Y}) \text{ and } |G\varphi| < R_1,$$

where $R_1 = C_6\mathcal{M}(R)$.

Since $\mathcal{M}(R)/R \to 0$ as $R \to 0$, and since $\|y\| \leq C_7|y|$ holds for some constant C_7 and all $y \in \mathbb{Y}$, it follows that there exists $\rho > 0$ such that for all $R \geq \rho$, we have

$$G(B_{AA(\mathbb{X})}(0, R)) \subset B_{BC^\delta(\mathbb{R}, \mathbb{Y})}(0, R) \cap B_{AA(\mathbb{X})}(0, R). \qquad (2.25)$$

Since G leaves $AA(\mathbb{X}) \subset BC(\mathbb{R}, \mathbb{X})$ invariant, the estimate (2.25) along with the continuity properties of G imply that G is a continuous, compact mapping $S \to S$, where S is the ball of radius R in $AA(\mathbb{X})$ and $R \geq \rho$.

By the Schauder fixed point theorem G has a fixed point in S, φ_0.

Obviously, φ_0 is a mild solution of (2.9).

Finally, the above results can be summarized as follows.

Theorem 2.23. *Consider the evolution equation (2.9) where A generates an exponentially stable C_0-semigroup $\mathcal{T}$ in $\mathcal{B}(\mathbb{X})$. Assume assumptions (2.10) and (2.14)-(2.17).*

Then (2.9) has a mild solution in $AA(\mathbb{X})$.

Now we give the following example.

Example of Nonuniqueness of Almost Automorphic Solutions

Let $\mathbb{X} = \mathbb{R}$, $A = -1$ and

$$u(t) = \begin{cases} t^{3/2}e^{1-t} & \text{for } t \in [0, \frac{3}{2}], \\ 0 & \text{for } t \in [-\frac{3}{2}, 0]. \end{cases}$$

Then for $t \in [0, \frac{3}{2}]$ we have

$$u'(t) = -u(t) + \frac{3}{2}t^{1/2}e^{(1-t)} = -u(t) + \frac{3}{2}u(t)^{1/3}e^{\frac{2}{3}(1-t)}$$

which can be written as

$$u'(t) = -u(t) + f(t, u(t))$$

where

$$f(t,\varphi) = \begin{cases} \frac{3}{2}\varphi^{1/3}e^{\frac{2}{3}(1-t)} & \text{for } t \in [0, \frac{3}{2}] \times \mathbb{R}, \\ \frac{3}{2}\varphi^{1/3}e^{2/3} & \text{for } t \in [-\frac{3}{2}, 0] \times \mathbb{R}. \end{cases}$$

Note that

$u'(\frac{3}{2}) = 0$ and $u(\frac{3}{2}) = (\frac{3}{2})^{\frac{3}{2}}e^{-\frac{3}{2}}$.

Now let

$$f(t,\varphi) = \begin{cases} f(\frac{3}{2}, \varphi) & \text{on } [\frac{3}{2}, 3] \times \mathbb{R}, \\ f(\frac{9}{2} - t, \varphi) & \text{on } [3, \frac{9}{2}] \times \mathbb{R}. \end{cases}$$

$$u(t) = \begin{cases} u(\frac{3}{2}) & \text{on } [\frac{3}{2}, 3], \\ u(\frac{9}{2} - t) & \text{on } [3, \frac{9}{2}], \\ 0, & \text{for } t = 0. \end{cases}$$

Extend u to be a periodic function of period 6 (hence an almost automorphic function).

Then u and $v \equiv 0$, both satisfy

$$\frac{dx}{dt} = -x + f(t, x), \quad x(0) = 0.$$

2.3 Optimal weak-almost periodic solutions

In this section we still consider equation (2.1) in a uniformly convex Banach space $(X, \|\cdot\|)$. We will prove existence and uniqueness of the so-called "optimal" almost periodic solutions in the weak sense, a result contained in [77].

We make the following assumptions:

A1: $A : D(A) \subset X \mapsto X$ is a linear operator (generally unbounded) that generates a C_0-semigroup of uniformly bounded linear operators $T(t)$, $t \in \mathbb{R}^+.$, i.e. there exists $M > 0$ such that $\sup_{t \in \mathbb{R}^+} \|T(t)\| = M$. For each $T(t)$, $t \in \mathbb{R}^+$, $T^*(t)$ will denote the adjoint operator of $T(t)$.

A2: $f : \mathbb{R} \mapsto X$ is a nontrivial strongly continuous function.

Now we denote by Ω_f, the set of all mild solutions $x(t)$ of equation (2.1) which are bounded over $\mathbb{R}$, that is

$$\mu(x) = \sup_{t \in \mathbb{R}} \|x(t)\| < \infty.$$

We assume here that $\Omega_f \neq \emptyset$.

Definition 2.24. *A bounded mild solution $\tilde{x}(t)$ of (2.1) will be called an optimal mild solution of (2.1) if*

$$\mu(\tilde{x}) \equiv \mu^* = \inf_{x \in \Omega_f} \mu(x).$$

Theorem 2.25. *Under assumptions A1-A2 and assuming that $\Omega_f \neq \emptyset$, equation (2.1) has a unique optimal mild solution.*

The proof is based on the following well-known fact (cf. [55], Corollary 8.2.1)

Lemma 2.26. *If K is a non-empty convex and closed subset of a uniformly convex Banach space X and $v \notin K$, then there exists a unique $k_0 \in K$ such that $\|v - k_0\| = \inf_{k \in K} \|v - k\|$.*

Now we begin the proof of the theorem:

Proof. Since the trivial solution $0 \notin \Omega_f$, it suffices to prove that Ω_f is a convex and closed set; we then use *Lemma 2.26* to deduce uniqueness of the optimal mild solution to equation (2.1). Let us first obtain convexity of Ω_f:

Consider two distinct bounded mild solutions $x_1(t)$ and $x_2(t)$, and a real number $0 \leq \lambda \leq 1$ and let $x(t) = \lambda x_1(t) + (1 - \lambda)x_2(t)$, $t \in \mathbb{R}$. $x(t)$ is continuous and has the integral representation

$$x(t) = T(t - t_0)x(t_0) + \int_{t_0}^{t} T(t - s)f(s)ds$$

for every $t_0 \in \mathbb{R}$ and $t \geq t_0$. Here we have

$$x(t_0) = \lambda x_1(t_0) + (1 - \lambda)x_2(t_0).$$

$x(t)$ is then a mild solution of equation (2.1). It is easy to see that $x(t)$ is bounded over $\mathbb{R}$ since

$$\mu(x) = \sup_{t \in \mathbb{R}} ||x(t)|| \leq \lambda \mu(x_1) + (1 - \lambda)\mu(x_2) < \infty.$$

We conclude that $x(t) \in \Omega_f$.

Let us now show that Ω_f is closed:

Consider a sequence (x_n) in Ω_f such that $\lim_{n \to \infty} x_n(t) = x(t)$, $t \in \mathbb{R}$. We have for all $t_0 \in \mathbb{R}$ and $t \geq t_0$:

$$x_n(t) = T(t - t_0)x_n(t_0) + \int_{t_0}^{t} T(t - s)f(s)ds$$

$n = 1, 2, \ldots$

For every fixed t and t_0 with $t \geq t_0$, we have

$$\lim_{n\to\infty} T(t-t_0)x_n(t_0) = T(t-t_0)\lim_{n\to\infty} x_n(t_0)$$
$$= T(t-t_0)x(t_0)$$

using continuity of the operator $T(t-t_0)$. We then deduce that

$$x(t) = T(t-t_0)x(t_0) + \int_{t_0}^{t} T(t-s)f(s)ds$$

for all $t_0 \in \mathbb{R}$, $t \geq t_0$. This shows that $x(t)$ is a mild solution of (2.1).

Finally we claim that $x(t)$ is bounded over $\mathbb{R}$. Indeed, let us write $x(t)$ as follows:

$$x(t) = T(t-t_0)x(t_0) + \int_{t_0}^{t} T(t-s)f(s)ds - x_n(t) + x_n(t)$$
$$= T(t-t_0)(x(t_0) - x_n(t_0)) + x_n(t)$$

for every $n = 1, 2, \ldots$, and every $t_0 \in \mathbb{R}$ and $t \geq t_0$. This gives

$$\|x(t)\| \leq M\|x(t_0) - x_n(t_0)\| + \|x_n(t)\|$$
$$\leq M\|x(t_0) - x_n(t_0)\| + \mu(x_n).$$

Choose n large enough such that $\|x(t_0) - x_n(t_0)\| \leq 1$. Then we have

$$\mu(x) \leq M + \mu(x_n) < \infty.$$

We have just proved that $x \in \Omega_f$, which completes the proof of the theorem. $\square$

Theorem 2.27. *Let $f \in AP(X)$ and assume that A1 and A2 hold true.*

Then the optimal mild solution of equation (2.1) is weakly almost periodic.

Proof. Consider $x(t)$ the optimal mild solution of equation (2.1). Such $x(t)$ is unique by the previous Theorem. We have

$$x(t) = T(t - t_0)x(t_0) + \int_{t_0}^t T(t - x)f(s)ds$$

for all $t_0 \in \mathbb{R}$, $t \geq t_0$. Let (s'_n) be an arbitrary sequence of real numbers. Since f is almost periodic, we can extract a subsequence $(s_n) \subset (s'_n)$ such that

$$\lim_{n \to \infty} f(t + s_n) = g(t)$$

uniformly in $t \in \mathbb{R}$. This fact is assured by Bochner's Criterion.

We know $g(t)$ is also strongly continuous. Now, for fixed $t_0 \in \mathbb{R}$, we can obtain a subsequence of (s_n), which again we will denote (s_n), such that

$$\mathrm{w} - \lim_{n \to \infty} x(t_0 + s_n) = v_0 \in X,$$

since X is a reflexive Banach space. The function

$$y(t) = T(t - t_0)v_0 + \int_{t_0}^t T(t - s)g(s)ds$$

is then strongly continuous. It is a mild solution of

$$x'(t) = Ax(t) + g(t), \quad t \in \mathbb{R}.$$

We can prove.
Lemma 2.28.

$$w - \lim_{n \to \infty} x(t + s_n) = y(t)$$

for each $t \in \mathbb{R}$.

Proof. Let us write

$$x(t + s_n) = T(t - t_0)x(t_0 + s_n) + \int_{t_0}^{t} T(t - s)f(s + s_n)ds$$

$n = 1, 2, \ldots$. Take x^* in X^* and let

$$< x^*, T(t - t_0)x(t_0 + s_n) > - < x^*, T(t - t_0)v_0 >$$

$$=< T^*(t - t_0)x^*, x(t_0 + s_n) - v_0 >$$

for every $n = 1, 2, \ldots$, from which we observe that the sequence $(T(t - t_0)x(t_0 + s_n))$ converges to $T(t - t_0)v_0$ in the weak sense.

Also we have the following inequalities

$$\int_{t_0}^{t} T(t - s)f(s + s_n)ds - \int_{t_0}^{t} T(t - s)g(s)ds$$

$$= \|\int_{t_0}^{t} T(t - s)(f(s + s_n) - g(s))ds\|$$

$$\leq \int_{t_0}^{t} \|T(t - s)\| \cdot \|f(s + s_n) - g(s)\|ds$$

$$\leq M \int_{t_0}^{t} \|f(s + s_n) - g(s)\|ds.$$

Therefore

$$\lim_{n \to \infty} \int_{t_0}^{t} T(t - s)f(s + s_n)ds = \int_{t_0}^{t} T(t - s)g(s)ds$$

in the strong sense and consequently in the weak sense in X. This proves the *Lemma* □

We also have:

Lemma 2.29. $\mu(y) = \mu(x) = \mu^*$

Proof. Since $x(t)$ is an optimal mild solution of equation (2.1), we have

$$\mu^* = \mu(x) = \sup_{t \in \mathbb{R}} \|x(t)\|.$$

Take an arbitrary x^* in X^*; then by *Lemma 2.28* we obtain

$$\lim_{n \to \infty} < x^*, x(t + s_n) > = < x^*, y(t) >$$

for every $t \in \mathbb{R}$. Now for each $n = 1, 2, \ldots$, we have

$$| < x^*, x(t + s_n) > | \leq \|x^*\| \|x(t + s_n)\| \leq \|x^*\| \mu^*.$$

Therefore, $| < x^*, y(t) > | \leq \|x^*\| \mu^*$, for every $t \in \mathbb{R}$, and consequently $\|y(t))\| \leq \mu^*$, for every $t \in \mathbb{R}$, so that $\mu(y) < \mu^*$.

Suppose that $\mu(y) < \mu^*$. Remark that $\lim_{n \to \infty} g(t - s_n) = f(t)$ uniformly in $t \in \mathbb{R}$ since $f \in AP(X)$. Also since X is a reflexive Banach space, we can extract from the sequence (s_n), a subsequence which we still denote (s_n) such that $(y(t_0 - s_n))$ is weakly convergent, say to $z \in X$. Now we have

$$\lim_{n \to \infty} y(t - s_n) = T(t - t_0)z + \int_{t_0}^t T(t - s)f(s)ds$$

in the weak sense, for every $t \in \mathbb{R}$. Let us consider the function

$$z(t) = T(t - t_0)z + \int_{t_0}^t T(t - s)f(s)ds.$$

It is a bounded mild solution of equation (2.1). For the same reasons stated above, we have

$$\mu(z) \leq \mu(y),$$

therefore $\mu(z) < \mu^*$, which is absurd by definition of μ^*.

We also need the following:
Lemma 2.30.

$$\mu(y) = \inf_{v \in \Omega_g} \mu(v)$$

i.e. $y(t)$ is an optimal mild solution of the equation

$$x'(t) = Ax(t) + g(t), \quad t \in \mathbb{R}.$$

Proof. By *Lemma 2.29*, $y(t)$ is bounded over $\mathbb{R}$. We know also that $y(t)$ is a mild solution of $x'(t) = Ax(t) + g(t)$, $t \in \mathbb{R}$. So $y(t) \in \Omega_g$.

It remains to prove that $y(t)$ is optimal.

Suppose it is not. Since $\Omega_g \neq \emptyset$, there exists a unique optimal solution $v(t)$ of

$$x'(t) = Ax(t) + g(t)$$

by *Theorem 2.25*. And $\mu(v) < \mu(y)$ and

$$v(t) = T(t - t_0)v(t_0) + \int_{t_0}^{t} T(t - s)g(s)ds$$

for all $t_0 \in \mathbb{R}$, $t \geq t_0$.

We can find a subsequence $(s_{n_k}) \subset (s_n)$ such that

$$\text{weak} - \lim_{k \to \infty} v(t - s_{n_k}) = T(t - t_0)z + \int_{t_0}^{t} T(t - s)f(s)ds$$
$$:= V(t)$$

Observe that $V(t) \in \Omega_f$ and

$$\mu(V) \leq \mu(v) < \mu(y)$$

which is absurd.

Therefore $y(t)$ is an optimal mild solution of

$$x'(t) = Ax(t) + g(t), \quad t \in \mathbb{R},$$

and in fact the only one by *Theorem 2.25*. □

Proof of Theorem 2.27(continued): To show that $x(t)$ is weakly almost periodic, it suffices to prove now that

$$weak - \lim_{n \to \infty} x(t + s_n) = y(t)$$

uniformly in $t \in \mathbb{R}$.

Suppose that it is not the case; then there exists $x^* \in X^*$ such that

$$\lim_{n \to \infty} < x^*, x(t + s_n) >=< x^*, y(t) >$$

is not uniform in $t \in \mathbb{R}$. Consequently, we can find a number $\alpha > 0$, a sequence (t_k) with two subsequences (s_k') and $(s_k")$ of (s_n) such that

$$| < x^*, x(t + s_k') - x(t + s_k") > | > \alpha \qquad (2.26)$$

for all $k = 1, 2, \ldots$

Let us again extract two subsequences of (s_n') and (s_n'') respectively, without changing the notation, such that

$$\lim_{k \to \infty} f(t + t_k + s_k') = g_1(t)$$

and

$$\lim_{k \to \infty} f(t + t_k + s_k'') = g_2(t)$$

both uniformly in $t \in \mathbb{R}$, since $f \in AP(X)$. As we did previously, we may obtain

$$weak - \lim_{k \to \infty} f(t + t_k + s_k') = T(t - t_0)z_1 + \int_{t_0}^{t} T(t - s)g_1(s)ds \equiv y_1(t)$$

and

$$\text{weak}-\lim_{k\to\infty} f(t+t_k+s_k'') = T(t-t_0)z_2 + \int_{t_0}^{t} T(t-s)g_2(s)ds \equiv y_2(t)$$

for each $t \in \mathbb{R}$, where $y_1(t)$ and $y_2(t)$ are optimal mild solutions in Ω_{g_1} and Ω_{g_2}, respectively.

Now we can show that $g_1(s) = g_2(s)$, $s \in \mathbb{R}$; indeed since $\lim_{k\to\infty} f(t + t_k + s_k)$ exists uniformly in $t \in \mathbb{R}$ and (s_k'), (s_k'') are two subsequences of (s_k), we will get

$$\sup_{s\in\mathbb{R}} \|f(s + s_k') - f(s + s_k'')\| < \varepsilon$$

if $k \geq k_0(\varepsilon)$ and consequently

$$\sup_{s\in\mathbb{R}} \|f(t + t_k + s_k') - f(t + t_k + s_k'')\| < \varepsilon$$

for $k \geq k_0(\varepsilon)$, which shows that $g_1(s) = g_2(s)$ for all $s \in \mathbb{R}$. By the uniqueness of the optimal mild solution we get $y_1(t) = y_2(t)$, $t \in \mathbb{R}$. But

$$y_1(0) = \text{weak} - \lim_{k\to\infty} x(t_k + s_k')$$

and

$$y_2(0) = \text{weak} - \lim_{k\to\infty} x(t_k + s_k'').$$

It is then clear that the equality $y_1(0) = y_2(0)$ contradicts the inequality (2.24) above and establishes the proof of the Theorem.
□

2.4 Existence of Weakly Almost Automorphic Solutions

We give in this section a result on the existence of a weakly almost automorphic solution to the equation (2.1). It is a slightly different version of a result in [101].

Theorem 2.31. *Let $(X, \| \cdot \|)$ be a reflexive, separable Banach space, and assume that A is the infinitesimal generator of a C_0-semigroup $(T(t))_{t \geq 0}$. Let X^* be the dual space of X and $T^*(t) \in L(X^*)$ the adjoint operator of $T(t)$, for each $t \geq 0$ with the property that*

$$\lim_{t \longrightarrow \infty} T^*(t)\varphi = 0 \text{ for every } \quad \varphi \in X^*$$

in the uniform operator topology. Assume also that f is weakly almost automorphic.

Then every bounded mild solution of (2.1) is weakly almost automorphic.

We first state and prove the following:

Lemma 2.32. *Under assumptions of the theorem, we claim that the functions $T(t-s)f(s)$, $T(t-s)g(s) : [a, t] \mapsto X$ are strongly measurable and $\|T(t-s)f(s)\|$, $\|T(t-s)g(s)\|$ are Lebesgue integrable.*

Proof. By strong continuity of $T(t-s)$ and weak continuity of $f(s)$, it is clear that $T(t-s)f(s)$ is weakly continuous, thus strongly measurable.

Moreover the set $B = \{T(t-s)f(s)/ \ s \in [a, t]\}$ is contained in the least closed subspace spanned by the set

$$\{T(t-s)f(s)/ \ s \in \mathbb{Q} \bigcap [a, t]\}$$

($\mathbb{Q}$ denotes the set of rational numbers).

Hence $\{T(t-s)f(s)/ \ s \in [a, t]\}$ is separable.

Also note that $T(t-s)g(s)$ is weakly measurable as pointwise limit of the following sequence of strongly measurable functions $T(t-s)f(s+s_n)$.

And since the Banach space X is assumed to be separable, strong and weak measurabilty are equivalent. Measurabilty of both numerical functions $\|T(t-s)f(s)\|$ and $\|T(t-s)g(s)\|$ is also easy to establish.

As a result of the lemma, the functions

$$T(t-s)f(s), \ T(t-s)g(s) : [a,t] \mapsto X$$

are integrable in Bochner's sense.

We are now ready to prove the theorem.

Proof. Let $x(t) = T(t-a)x(a) + \int_a^t T(t-s)f(s)ds \ \ t \geq a$ be a mild solution of the equation (2.1) such that $sup_{t\in\mathbb{R}}\|x(t)\| = M < \infty$. Given an arbitrary sequence of real numbers (s_n'), consider the functions $x_n(t)$ defined by

$$x_n(t) := x(t + s_n') \ \ t \in \mathbb{R}.$$

Since for each $t \in \mathbb{R}$, the sequence $(x_n(t))$ is bounded, there exists a subsequence $(s_{n,0})$ of (s_n') such that

$$w - \lim_{n\mapsto\infty} x_{n,0}(0) = w - \lim_{n\mapsto\infty} x(s_{n,0}) = y_0$$

exists in X by *Proposition 1.2.18* in [80]. From the sequence $(s_{n,0})$, we can extract a subseqence $(s_{n,1})$ such that

$$w - \lim_{n\mapsto\infty} x_{n,1}(-1) = w - \lim_{n\mapsto\infty} x(-1 + s_{n,1}) = y_1$$

exists in X. We continue the process inductively and we take the diagonal sequence (s_n) to obtain

$$w - \lim_{n \mapsto \infty} x_n(-N) = w - \lim_{n \mapsto \infty} x(-N + s_n) = y_N, \quad N = 0, 1, 2, ...$$

Now using weak almost automorphy of the function f, we can find a subsequence of (s_n) which denote again by (s_n) such that

$$w - \lim_{n \mapsto \infty} x(-N + s_n) = y_N, \quad N = 0, 1, 2, ...$$

$$w - \lim_{n \mapsto \infty} f(t + s_n) = g(t) \text{ for each } t \in \mathbb{R}$$

$$w - \lim_{n \mapsto \infty} g(t - s_n) = f(t) \text{ for each } t \in \mathbb{R}.$$

We now need to prove that

$$w - \lim_{n \mapsto \infty} x(t + s_n) \text{ exists for each } t \in \mathbb{R}.$$

Fix $t \in \mathbb{R}$ and choose N such that $-N < t$. Then it is

$$x(t + s_n) = T(t + N)x(-N + s_n) + \int_{-N}^{t} T(t - s)f(s + s_n)ds.$$

Take an arbitrary $\varphi \in X^*$. Then using the duality $\langle \cdot, \cdot \rangle$ between X and X^*, we get

$$\langle \varphi, \int_{-N}^{t} T(t - s)f(s + s_n) - g(s)ds \rangle =$$

$$\int_{-N}^{t} \langle T^*(t - s)\varphi, f(s + s_n) - g(s) \rangle ds.$$

Now we put

$$F_n(s) = \langle \varphi, T(t - s)f(s + s_n) - g(s) \rangle, \quad n = 1, 2, ...$$

From the lemma above it follows immediately that $(F_n(s))$ is a sequence of measurable functions defined on the compact interval $[-N, t]$. Moreover we have the inequality

$$|F_n(s)| \leq \|\varphi\| \|T(t-s)\| (\|f(s+s_n) + g(s)\|).$$

Since $\|T(t)\| \leq Ke^{\omega t}$ for every $t \geq 0$, the bound $\|T(t-s)\| \leq L$ holds true on $[-N, t]$ for some $0 < L < \infty$. Note also that both $f(t)$ and $g(t)$ are bounded (cf *Proposition 1.32*).

Hence $(F_n(s))$ is a bounded sequence of measurable functions on $[-N, t]$. Also it is clear that $w - \lim_{n \mapsto \infty} F(s) = 0$ everywhere on $[-N, t]$; thus

$$\lim_{n \mapsto \infty} \int_{-N}^{t} F(s)ds = 0,$$

that is

$$w - \lim_{n \mapsto \infty} \int_{-N}^{t} T(t-s)f(s+s_n)ds = \int_{-N}^{t} T(t-s)g(s)ds.$$

We infer that $w - \lim_{n \mapsto \infty} x(t+s_n) = y(t)$ where

$$y(t) = T(t+N)y(-N) + \int_{-N}^{t} T(t-s)g(s)ds.$$

Now take any pair of real numbers a, t such that $t > a$ and choose a positive integer N such that $-N < a < t$. Then we get

$$y(t) = T(t-a)y(a) + \int_{a}^{t} T(t-s)g(s)ds.$$

We have also

$$\|y(t)\| \leq \liminf_{n \mapsto \infty} \|x(t+s_n)\| \leq M \quad \text{for every} \quad t \in \mathbb{R}.$$

Hence

$$\sup_{t \in \mathbb{R}} \|y(t)\| \leq M.$$

Repeating the argument that we used to show the existence of $w-\lim_{n \mapsto \infty} x(t + s_n)$, we can show that there exists a subsequence of (s_n) which we still designate by (s_n) such that $w-\lim_{n \mapsto \infty} y(t - s_n)$ exists for every $t \in \mathbb{R}$.

Call $z(t)$ this limit, then we have

$$z(t) = T(t - a)z(a) + \int_a^t T(t - s)f(s)ds,$$

with

$$\sup_{t \in \mathbb{R}} \|z(t)\| \leq \sup_{t \in \mathbb{R}} \|y(t)\| \leq M.$$

Finally let us show that $z(t) = x(t)$ for every $t \in \mathbb{R}$. We fix $t \in \mathbb{R}$ and choose arbitrary $\varepsilon > 0$ and $\varphi \in X^*$. We have

$$|\langle \varphi, z(t) - x(t) \rangle| = |\langle \varphi, T(t - a)z(a) - x(a) \rangle|$$

$$= |\langle T^*(t - a)\varphi, z(a) - x(a) \rangle| \leq 2M\|T^*(t - a)\varphi\|$$

Since $s - \lim_{n \mapsto \infty} T^*(t)\varphi = 0$ for every $\varphi \in X^*$, we can choose a negative number a sufficiently large so that $\|T^*(t - a)\varphi\| < \frac{\varepsilon}{2M}$; which gives

$$|\langle \varphi, z(t) - x(t) \rangle| = 0 \text{ for every } \varphi \in X^*,$$

that means

$$z(t) = x(t) \text{ for each } t \in \mathbb{R}.$$

The proof is now established. $\square$

2.5 A Correspondence Between Linear and Nonlinear Equations

Consider in a Banach space $(X, \|.\|)$ the differential equation

$$x'(t) = Ax(t) + f(t), \quad t \in \mathbb{R}, \tag{2.27}$$

and the nonlinear differential equation

$$x'(t) = Ax(t) + f(t) + g(t, x(t)), \quad t \in \mathbb{R}, \tag{2.28}$$

where A is the infinitesimal generator of a C_0-group $T = (T(t))_{t \in \mathbb{R}}$ of bounded linear operators on X, such that $\sup_{t \in \mathbb{R}} \|T(t)\| = M < \infty$. Let $\Omega_1 :=$ the space of all bounded mild solutions of (2.27) and $\Omega_2 :=$ the space of all bounded mild solutions of (2.28).

Let $C_b(\mathbb{R}; X)$ denote the Banach space of continuous bounded functions $u : \mathbb{R} \to X$ equipped with the norm $\|u\|_0 = \sup_{t \in \mathbb{R}} \|u(t)\|$. Then obviously $\Omega_i \subset C_b(\mathbb{R}; X)$, for each $i = 1, 2$.

We begin with the following:

Theorem 2.33. *Assume that the function $f : \mathbb{R} \to X$ is (strongly) continuous and $g : \mathbb{R} \times X \to X$ is (strongly) jointly continuous in t and x. Let*

$$\|g(t, x) - g(t, y)\| \le h(t)\|x - y\|, \quad \text{for every} \quad t \in \mathbb{R}$$

and, $x, y \in X$ where the numerical function $h \in L^1(\mathbb{R})$ and

$$\int_{\mathbb{R}} h(t)\, dt < \frac{1}{M}.$$

We assume also that

$$\int_{\mathbb{R}} \|g(t,0)\| \, dt < \infty.$$

Then Ω_1 and Ω_2 are homeomorphic.

Proof: Consider the mapping $F : C_b(\mathbb{R}; X) \longrightarrow C_b(\mathbb{R}; X)$ defined by

$$Fx(t) = \int_0^t T(t-s)g(s, x(s)) \, ds, \quad t \in \mathbb{R}$$

for each continuous bounded function $x(t) : \mathbb{R} \to X$. It is easy to observe that F is well-defined on $C_b(\mathbb{R}; X)$. Indeed we have

$$\|Fx(t)\| \le M \int_0^t \|g(s, x(s))\| \, ds$$

$$\le M \left[\int_0^t \|g(s, x(s)) - g(s,0)\| \, ds + \int_0^t \|g(s,0)\| \, ds \right]$$

$$\le M \left[\int_0^t h(s)\|x(s)\| \, ds + \int_0^t \|g(s,0)\| \, ds \right]$$

$$\le M \left[K \int_{\mathbb{R}} h(s) \, ds + \int_{\mathbb{R}} \|g(s,0)\| \, ds \right]$$

$$< \infty$$

for each $t \in \mathbb{R}$, where $K = \|x\|_0 = \sup_{t \in \mathbb{R}} \|x(t)\|$.

Also for each pair, $u, v \in C_b(\mathbb{R}, X)$, we have

$$\|Fu(t) - Fv(t)\| = \left\| \int_0^t T(t-s)\Big(g(s, u(s)) - g(s, v(s))\Big) \, ds \right\|$$

$$\le M\|u - v\|_0 \int_0^t h(s) \, ds$$

$$\le c\|u - v\|_0$$

where the constant $c = M \int_{\mathbb{R}} h(t) \, dt$. So the mapping F is a strict contraction.

Now let $z(t)$ be a bounded mild solution of (2.27) with the representation

$$z(t) = T(t)z(0) + \int_0^t T(t-s)f(s)\,ds, \quad t \in \mathbb{R}.$$

Consider the mapping $S : C_b(\mathbb{R}; X) \longrightarrow C_b(\mathbb{R}; X)$ defined by

$$Su(t) = z(t) + Fu(t)$$

for each $u(t) \in C_b(\mathbb{R}; X)$.

S is also a strict contraction on $C_b(\mathbb{R}; X)$. Hence it possesses a unique fixed point, say $w(t)$, which satisfies the equation

$$w(t) = Sw(t) = z(t) + Fw(t)$$

for each $t \in \mathbb{R}$. That is,

$$
\begin{aligned}
w(t) &= T(t)z(0) + \int_0^t T(t-s)\Big(f(s) + g(s, w(s))\Big)\,ds \\
&= T(t)w(0) + \int_0^t T(t-s)\Big(f(s) + g(s, w(s))\Big)\,ds
\end{aligned}
$$

for each $t \in \mathbb{R}$, since $z(0) = w(0)$.

Obviously, $w(t)$ is a mild solution of (2.28). It is bounded as the sum of two bounded functions and is obviously continuous.

On the other hand, let

$$x(t) = T(t)x(0) + \int_0^t T(t-s)\Big(f(s) + g(s, w(s))\Big)\,ds$$

be a given bounded mild solution of (2.28), and consider

$$z(t) = T(t)x(0) + \int_0^t T(t-s)f(s)\,ds.$$

Then $z(t)$ is a bounded mild solution of (2.27) and $z(t) = x(0)$, $z(t) = x(t) - Fx(t)$, $t \in \mathbb{R}$. The fact that the mapping $z \longmapsto x$ is a homeomorphism follows from the inequalities

$$\|x_1 - x_2\|_0 \le \frac{1}{1-c}\|z_1 - z_2\|_0$$

and

$$\|z_1 - z_2\| \le (1+c)\|x_1 - x_2\|_0.$$

The theorem is proved. □

Theorem 2.34. *Let the functions f and g have the properties described in Theorem 2.33. Assume in addition that $f \in AA(X)$.*

Then every mild solution of (2.28) restricted to $\mathbb{R}^+$ is asymptotically almost automorphic.

Proof: Let $x(t)$ be a mild solution of (2.28) restricted to $\mathbb{R}^+$. So

$$x(t) = T(t)x(0) + \int_0^t T(t-s)\Big(f(s) + g(s, x(s))\Big)\, ds$$

for each $t \in \mathbb{R}^+$. Observe that the function

$$z(t) = T(t)x(0) + \int_0^t T(t-s)f(s)\, ds, \quad t \in \mathbb{R}$$

is almost automorphic.

Writing $g(s, x(s)) = g(s, x(s)) - g(s, 0) + g(s, 0)$, we obtain the inequality

$$\left\| \int_0^\infty T(-s)g(s, x(s))ds \right\|$$
$$\le M \left(K \int_0^\infty h(s)ds + \int_0^\infty \|g(s, 0)\|ds \right) < \infty,$$

which shows that the improper integral

$$\int_0^\infty T(-s)g(s, x(s))ds \quad \text{exists in } X.$$

Consequently the function $\mathbb{R} \to X$ defined by

$$T(t) \int_0^\infty T(-s)g(s, x(s))ds = \int_0^\infty T(t-s)g(s, x(s))\, ds$$

is almost automorphic. On the other hand, for each $t \in \mathbb{R}^+$ we have

$$\left\| \int_t^\infty T(t-s)g(s, x(s))ds \right\| \le M(K \int_t^\infty h(s)ds$$
$$+ \int_t^\infty \|g(s, 0)\|ds$$

which shows that

$$\lim_{t \to \infty} \left\| \int_t^\infty T(t-s)g(s, x(s))ds \right\| = 0.$$

Finally, if we write

$$x(t) = z(t) + \int_0^\infty T(t-s)g(s, x(s))ds - \int_t^\infty T(t-s)g(s, x(s))ds$$

for each $t \in \mathbb{R}^+$, we see that $x(t)$ is indeed asymptotically almost automorphic. $\square$

We now state the following theorem whose proof is a combination of the above results:

Theorem 2.35. *Under the assumption of Theorem 2.33 and Theorem 2.34, the spaces $AA(X) \bigcap \Omega_1$ and $AAA(X) \bigcap \Omega_2$ are homeomorphic.*

Bibliographical Remarks and Open Problems

Among important results in this chapter, we like to mention
N'Guérékata's contribution (*Theorem 2.1*) and the so-called method
of reduction (*Theorem 2.4*) also refered as a result of Bohr-
Neugebauer type which has various generalizations for instance in
[62], and other cases in [76] where the use of a method of decomposition of the space has been successful. In perturbed equations
of the form

$$x'(t) = (A + B)x(t)$$

and its purturbation

$$x'(t) = (A + B)x(t) + f(t),$$

where both operators A and B are unbounded, the use of the
invariant subspaces theory in [31] has successfully produced almost
automorphy of solutions. *Theorem 2.15* is contanied in [28].

For nonlinear equations, a partial result on the existence and
the uniqueness of almost automorphic solutions was obtained in
Theorem 2.17 (see also [75]) when A generates an exponentially
stable semigroup. Its variant, *Theorem 2.23*, is a contribution by J.
A. Goldstein and G.M. N'Guérékata (see [46]).This result is new,
even for the almost periodic case.

Another important result is contained in [32], in the case of
holomorphic semigroup, using the method of sums of commut-
ing operators. The same paper introduces the notion of uniform
spectrum of the the forced term f. The ultimate aim of this pio-
neering work is to compare the spectrum of f and the solution's

one as in almost periodic case. This work has generated several other papers, and the method introduced there can be applied to higher order evolution equations (see for instance [58]). It is also of great interest to study linear inhomogeneous evolution equations of the for $x' = Ax + f$ where the operator A generates instead a C-semigroup.

For linear functional differential equations, Y. Hino and S. Murakami studied the existence of almost aunorphic solutions in their interesting paper [50]. One might obtain an easier proof of their results using the methods presented in this chapter.

Theorem 2.35 is a variant of *Theorem 6.1.3* in [80].

3

Almost Periodicity in Fuzzy Setting

In this chapter, the theory of almost periodicity as known in Banach spaces (see for instance [2], [23], [80]), is studied in fuzzy setting. It is based on a work by B. Bede and S. G. Gal [40]. We start with a brief overview of basic properties of the so-called fuzzy sets in *Section 3.1*. Then we introduce the notion of almost periodic fuzzy functions in *Section 3.2*, and study their harmonics in *Section 3.3*. Integration and differentiability of fuzzy functions are also introduced. Finally we apply the results obtained to fuzzy differential equations in *Section 3.4*.

3.1 Fuzzy Sets

Definition 3.1. *Given a set $X \neq \emptyset$, a fuzzy subset of X is a mapping $u : X \to [0, 1]$ and obviously any classical subset A of X can be considered as a fuzzy subset of X defined by $\chi_A : X \to [0, 1]$, $\chi_A(x) = 1$, if $x \in A$, $\chi_A(x) = 0$ if $x \in X \setminus A$.*

Definition 3.2. *Let us denote by* $\mathbb{R}_{\mathcal{F}}$ *the class of fuzzy subsets of the real axis* $\mathbb{R}$ *(i.e.* $u : \mathbb{R} \to [0,1]$*), satisfying the following properties:*

(i) $\forall u \in \mathbb{R}_{\mathcal{F}}$*, u is normal i.e.* $\exists x_u \in \mathbb{R}$ *with* $u(x_u) = 1$*;*

(ii) $\forall u \in \mathbb{R}_{\mathcal{F}}$*, u is convex fuzzy set, i.e.*

$$u\left(tx + (1-t)y\right) \geq \min\left\{u(x), u(y)\right\},$$

$\forall t \in [0,1]$*,* $x, y \in \mathbb{R}$*;*

(iii) $\forall u \in \mathbb{R}_{\mathcal{F}}$*, u is upper semi-continuous on* $\mathbb{R}$*;*

(iv) $\overline{\{x \in \mathbb{R} : u(x) > 0\}}$ *is compact.*

Then $\mathbb{R}_{\mathcal{F}}$ *is called the space of fuzzy real numbers.*

Remark 3.3. It is clear that $\mathbb{R} \subset \mathbb{R}_{\mathcal{F}}$, because any real number $x_0 \in \mathbb{R}$, can be described as the fuzzy number whose value is 1 for $x = x_0$ and 0 otherwise.

We will collect some other definitions and notations needed in the sequel. For $0 < r \leq 1$ and $u \in \mathbb{R}_{\mathcal{F}}$, we define

$$[u]^r := \{x \in \mathbb{R}; u(x) \geq r\}$$

$$[u]^0 := \overline{\{x \in \mathbb{R}; u(x) > 0\}}.$$

Now it is well known that for each $r \in [0,1]$, $[u]^r$ is a bounded closed interval. For $u, v \in \mathbb{R}_{\mathcal{F}}$ and $\lambda \in \mathbb{R}$, we have the sum $u \oplus v$ and the product $\lambda \odot u$ defined by

$$[u \oplus v]^r = [u]^r + [v]^r, \quad [\lambda \odot u]^r = \lambda [u]^r, \quad \forall r \in [0,1],$$

where $[u]^r + [v]^r$ means the usual addition of two intervals (as subsets of $\mathbb{R}$) and $\lambda [u]^r$ means the usual product between a scalar and a subset of $\mathbb{R}$. (see e.g. [20])

Now we define $D : \mathbb{R}_{\mathcal{F}} \times \mathbb{R}_{\mathcal{F}} \to \mathbb{R}_+ \cup \{0\}$ by

$$D(u,v) = \sup_{r \in [0,1]} \max \left\{ \left| u_-^r - v_-^r \right|, \left| u_+^r - v_+^r \right| \right\},$$

where $[u]^r = \left[u_-^r, u_+^r \right]$, $[v]^r = \left[v_-^r, v_+^r \right]$.

We also have the following well-known properties ([20]):

(a) $D(u \oplus w, v \oplus w) = D(u,v)$, $\forall u, v, w \in \mathbb{R}_{\mathcal{F}}$;

(b) $D(k \odot u, k \odot v) = |k| \, D(u,v)$, $\forall u, v \in \mathbb{R}_{\mathcal{F}}, \forall k \in \mathbb{R}$;

(c) $D(u \oplus v, w \oplus e) \leq D(u,w) + D(v,e)$, $\forall u, v, w, e \in \mathbb{R}_{\mathcal{F}}$ and $(\mathbb{R}_{\mathcal{F}}, D)$ is a complete metric space.

Also, the following is known:

Theorem 3.4. *(i) If we denote $\tilde{0} = \chi_{\{0\}}$ then $\tilde{0} \in \mathbb{R}_{\mathcal{F}}$ is neutral element with respect to $\oplus$, i.e. $u \oplus \tilde{0} = \tilde{0} \oplus u = u$, for all $u \in \mathbb{R}_{\mathcal{F}}$.*

(ii) With respect to $\tilde{0}$, none of $u \in \mathbb{R}_{\mathcal{F}} \backslash \mathbb{R}$ has opposite in $\mathbb{R}_{\mathcal{F}}$ (with respect to $\oplus$).

(iii) For any $a, b \in \mathbb{R}$ with $a, b \geq 0$ or $a, b \leq 0$, and any $u \in \mathbb{R}_{\mathcal{F}}$, we have

$$(a + b) \odot u = a \odot u \oplus b \odot u.$$

For general $a, b \in \mathbb{R}$, the above property does not hold.

(iv) For any $\lambda \in \mathbb{R}$ and any $u, v \in \mathbb{R}_{\mathcal{F}}$, we have

$$\lambda \odot (u \oplus v) = \lambda \odot u \oplus \lambda \odot v.$$

(v) For any $\lambda, \mu \in \mathbb{R}$ and any $u \in \mathbb{R}_{\mathcal{F}}$, we have

$$\lambda \odot (\mu \odot u) = (\lambda \cdot \mu) \odot u.$$

(vi) If we denote $\|u\|_{\mathcal{F}} = D(u, \tilde{0})$, $\forall u \in \mathbb{R}_{\mathcal{F}}$, then $\|\cdot\|_{\mathcal{F}}$ has the properties of a usual norm on $\mathbb{R}_{\mathcal{F}}$, i.e. $\|u\|_{\mathcal{F}} = 0$ if and only if $u = \tilde{0}$, $\|\lambda \odot u\|_{\mathcal{F}} = |\lambda| \cdot \|u\|_{\mathcal{F}}$ and

$$\|u \oplus v\|_{\mathcal{F}} \leq \|u\|_{\mathcal{F}} + \|v\|_{\mathcal{F}}, \quad |\|u\|_{\mathcal{F}} - \|v\|_{\mathcal{F}}| \leq D(u, v).$$

Remark 3.5. The properties (ii) and (iii) in *Theorem 3.4* show us that $(\mathbb{R}_{\mathcal{F}}, \oplus, \odot)$ is not a linear space over $\mathbb{R}$ and consequently $(\mathbb{R}_{\mathcal{F}}, \|\cdot\|_{\mathcal{F}})$ cannot be a normed space. However, the properties of D and those in *Theorem 3.4*, (iv)-(vi), have as an effect that most of the metric properties of a function defined on $\mathbb{R}$ with values in a Banach space, can be extended to functions $f : \mathbb{R} \to \mathbb{R}_{\mathcal{F}}$, called fuzzy functions.

3.2 Almost Periodicity in Fuzzy Setting

In this section , we present a fuzzy version of the theory of almost periodic functions as known in Banach spaces (see for instance [2], [23]), or Fréchet spaces (see [80]).

Definition 3.6. *A generalized fuzzy trigonometric polynomial of degree $\leq n$ is defined as a finite sum of the form $\sum^* T_k(x) \odot c_k$, where $c_k \in \mathbb{R}_{\mathcal{F}}$ and $T_k(x)$ are usual trigonometric polynomials of degree $\leq n$ i.e.*

$$T_k(x) = \sum_{j=0}^{n} [a_j^{(k)} \cos jx + b_j^{(k)} \sin jx], \quad a_j^{(k)}, b_j^{(k)} \in \mathbb{R}.$$

Here, the polynomials $T_k(x)$, are not necessarily supposed to be linearly independent on $\mathbb{R}$ and $\sum^*$ denotes addition (sum) with respect to $\oplus$ in $\mathbb{R}_{\mathcal{F}}$.

Note that the usual fuzzy trigonometric polynomials could naturally be defined of the form

$$\sum^* [a_j \odot \cos jx \oplus b_j \odot \sin jx],$$

with $a_j, b_j \in \mathbb{R}$, which because of the lack of distributivity of $\odot$ with respect to $\oplus$, obviously are only particular cases of the above concept of generalized fuzzy trigonometric polynomials. But, as it was pointed out in e.g. [3], the suitable fuzzy polynomials in approximation are of generalized type and not of usual type.

We also recall the following definition.

Definition 3.7. $f : \mathbb{R} \to \mathbb{R}_{\mathcal{F}}$ is said to be continuous at $x_0 \in \mathbb{R}$ if: $\forall \varepsilon > 0$, $\exists \delta > 0$ such that $D(f(x), f(x_0)) < \varepsilon$, whenever $x \in \mathbb{R}$, $|x - x_0| < \delta$.

Definition 3.8. Let $f : \mathbb{R} \to \mathbb{R}_{\mathcal{F}}$ be continuous on $\mathbb{R}$.

(i) We say that f is B-almost periodic if : $\forall \varepsilon > 0$, $\exists\, l > 0$ such that any interval of the form [a, a+l] contains at least a point τ with

$$D(f(t + \tau), f(t)) < \varepsilon, \forall t \in \mathbb{R}.$$

(ii) We say that f is normal if for any sequence $F_n : \mathbb{R} \to \mathbb{R}_{\mathcal{F}}$ of the form $F_n(x) = f(x + h_n)$, $n \in \mathbb{N}$, where $(h_n)_n$ is a sequence of real numbers, one can extract a subsequence of $(F_n)_n$, converging uniformly on $\mathbb{R}$ (i.e. $\forall (h_n)_n$, $\exists (h_{n_k})$, $\exists F : \mathbb{R} \to \mathbb{R}_{\mathcal{F}}$ (which may depend on $(h_n)_n$), such that $D\left(F_{n_k}(x), F(x)\right) \to 0$, as $k \to \infty$, uniformly with respect to $x \in \mathbb{R}$)

(iii) We say that f has the approximation property, if $\forall \varepsilon > 0$, exists some generalized fuzzy trigonometric polynomial T with $D\left(f(x), T(x)\right) < \varepsilon, \forall x \in \mathbb{R}$.

Remark 3.9. We observe that *Definition 3.8 i)* is compatible with *Definition 3.1.1* [80].

Also property ii) in *Definition 3.8* is the so-called *Bochner's criterion (Theorem 1.61)*.

Let us denote $AP\left(\mathbb{R}_{\mathcal{F}}\right) = \{f : \mathbb{R} \to \mathbb{R}_{\mathcal{F}}; f$ is B-almost periodic$\}$. We will show in the next two theorems that $AP\left(\mathbb{R}_{\mathcal{F}}\right)$ is a subclass of uniformly continuous bounded functions.

Theorem 3.10. *If $f : \mathbb{R} \to \mathbb{R}_{\mathcal{F}}$ is (continuous) B-almost periodic then f is bounded (i.e. $\exists M > 0$ with $D(f(x), f(y)) \le M,\ \forall x, y \in \mathbb{R}$).*

Proof. We follow the proof in [23, Theorem 6.1, p.154]. Because

$$D\left(f(x), f(y)\right) \le D\left(f(x), \tilde{0}\right) + D\left(\tilde{0}, f(y)\right) = \|f(x)\|_{\mathcal{F}} + \|f(y)\|_{\mathcal{F}},$$

it is sufficient to prove that $\exists M_1 > 0$ with $\|f(x)\|_{\mathcal{F}} < M_1$.

Let $\varepsilon = 1$ and $l(1)$ be as in *Definition 3.8*, (i). As in [23, Theorem 6.1, p.154] it follows $\|f(x)\|_{\mathcal{F}} \le M_1, \forall x \in [0, l(1)]$.

Now, if $t \in \mathbb{R}$ is arbitrary, then in $[-t, -t + l(1)]$ there is at least a point $\xi = \xi(1)$ in *Definition 3.8 i)*. Hence

$$\|f(t)\|_{\mathcal{F}} = D\left(f(t), \tilde{0}\right) = D\left(f(t) \oplus f(t + \xi), f(t + \xi) \oplus \tilde{0}\right)$$

$$\leq D\left(f\left(t\right),f\left(t+\xi\right)\right)+D\left(f\left(t+\xi\right),\tilde{0}\right)<1+M_{1},$$

because $t+\xi\in[0,l\left(1\right)]$, which proves the theorem. $\square$

Theorem 3.11. *If $f:\mathbb{R}\to\mathbb{R}_{\mathcal{F}}$ is B-almost periodic then f is uniformly continuous on $\mathbb{R}$.*

Proof. Following *Theorem 6.2, p. 154* in [23], and in view of the properties of D, we have

$$
\begin{aligned}
D\left(f\left(t_{2}\right),f\left(t_{1}\right)\right) &= D(f\left(t_{2}\right)\oplus f\left(t_{2}+\tau\right),f\left(t_{1}\right)\oplus f\left(t_{2}+\tau\right))\\
&= D(f\left(t_{2}\right)\oplus f\left(t_{2}+\tau\right),f\left(t_{2}+\tau\right)\oplus f\left(t_{1}\right))\\
&\leq D\left(f\left(t_{2}\right),f\left(t_{2}+\tau\right)\right)\\
&\quad + D\left(f\left(t_{1}\right),f\left(t_{2}+\tau\right)\right)\\
&= D\left(f\left(t_{2}\right),f\left(t_{2}+\tau\right)\right)+D(f\left(t_{1}\right)\\
&\quad \oplus f\left(t_{1}+\tau\right),f\left(t_{2}+\tau\right)\oplus f\left(t_{1}+\tau\right))\\
&\leq D\left(f\left(t_{2}\right),f\left(t_{2}+\tau\right)\right)+D\left(f\left(t_{1}\right),f\left(t_{1}+\tau\right)\right)\\
&\quad + D\left(f\left(t_{1}+\tau\right),f\left(t_{2}+\tau\right)\right).\square
\end{aligned}
$$

We also have:

Theorem 3.12. *If $f:\mathbb{R}\to\mathbb{R}_{\mathcal{F}}$ is B-almost periodic, then $\lambda\odot f$ $(\lambda\in\mathbb{R})$, $F_{h}\left(x\right)=f\left(x+h\right)$ $(x\in\mathbb{R})$ and $G\left(x\right)=\|f\left(x\right)\|_{\mathcal{F}}$, $(x\in\mathbb{R})$ are B-almost periodic.*

Proof. Because

$$D\left(\lambda\odot f\left(t+\xi\right),\lambda\odot f\left(t\right)\right)=|\lambda|\,D\left(f\left(t+\xi\right),f\left(t\right)\right),$$

for all $\lambda\in\mathbb{R}$, it easily follows that $\lambda\odot f$ is B-almost periodic.

And since

$$|\,\|f(t+\xi)\|_{\mathcal{F}} - \|f(t)\|_{\mathcal{F}}| \le D(f(t+\xi), f(t)),$$

then it is immediate that $G(x) = \|f(x)\|_{\mathcal{F}}$, $x \in \mathbb{R}$, is B-almost periodic (in the usual sense of function $G : \mathbb{R} \to \mathbb{R}$).

Let $h \in \mathbb{R}$ be fixed and for $\varepsilon > 0$, let $l > 0$ and τ attached to f in *Definition 3.8 i)*. By

$$D(f(t+\xi), f(t)) < \varepsilon, \quad \forall t \in \mathbb{R},$$

we get (by taking $t = u + h$)

$$D(f(u+h+\xi), f(u+h)) < \varepsilon, \quad \forall t \in \mathbb{R},$$

which immediately implies F_h is B-almost periodic.   □

The next result shows that $AP(\mathbb{R}_{\mathcal{F}})$ is closed with respect to uniform convergence on $\mathbb{R}$.

Theorem 3.13. *If $f_n : \mathbb{R} \to \mathbb{R}_{\mathcal{F}}$, $n \in \mathbb{N}$ are B-almost periodic and $f_n \to f$ as $n \to \infty$ uniformly on $\mathbb{R}$ (i.e. $\forall \varepsilon > 0$, $\exists n_0 \in \mathbb{N}$, such that*

$$D(f_n(x), f(x)) < \varepsilon, \quad \forall n \ge n_0, \quad \forall x \in \mathbb{R}),$$

then f is B-almost periodic.

Proof. It is similar to the proof of *Theorem 3.3* [23], or *Theorem 3.1.4* [80], by taking into account the properties of D in *Section 3.1*.   □

Theorem 3.14. *The set of values of $f : \mathbb{R} \to \mathbb{R}_{\mathcal{F}}$ supposed to be B-almost periodic, is relatively compact in the complete metric space $(\mathbb{R}_{\mathcal{F}}, D)$.*

Proof. We just follow the Proof of *Theorem 3.4* [23], or *Theorem 3.1.5* [80], with in mind the fact that in complete metric spaces, the relatively compact sets coincide with precompact sets, it is sufficient to show that for any $\varepsilon > 0$, the set of values of the function can be embedded in a finite number of spheres of radius ε. □

Remark 3.15. Let $f : \mathbb{R} \to \mathbb{R}_{\mathcal{F}}$ be B-almost periodic and let us consider the sequence of values $(f(t_n))_{n \in \mathbb{N}}$.

Denote $A = \{f(t_n) ; n \in \mathbb{N}\}$ and take the closure $\bar{A} \subset \overline{f(\mathbb{R})} \subset \mathbb{R}_{\mathcal{F}}$, it follows that $\bar{A}$ is compact, so $\bar{A}$ is sequentially compact too, which by $A \subset \bar{A}$ implies that the sequence $(f(t_n))_n$ has convergent subsequence in $\mathbb{R}_{\mathcal{F}}$.

The following result shows that the concepts in *Definition 3.8* i), and (ii), in fact are equivalent.

Theorem 3.16. *A function $f : \mathbb{R} \to \mathbb{R}_{\mathcal{F}}$ is B-almost periodic if and only if it is normal.*

Proof. It is similar to the proof of *Theorem 6.6, p. 156* in [23]. □

Also, we have

Theorem 3.17. *The sum $\oplus$, of two B-almost periodic functions is B-almost periodic.*

Proof. Similar to the proof of *Theorem 6.7* in [23]. □

Theorem 3.18. *If $f : \mathbb{R} \to \mathbb{R}_{\mathcal{F}}$ has the approximation property in Definition 3.8 (iii), then f is B-almost periodic.*

Proof. A function $f : \mathbb{R} \to \mathbb{R}_{\mathcal{F}}$ is called s_0-periodic if $f(t + s_0) = f(t)$, $\forall t \in \mathbb{R}$. Obviously a s_0-periodic function is B-almost periodic.

It follows by *Theorem 3.12* and *Theorem 3.17* above that any generalized fuzzy trigonometric polynomial is B-almost periodic, which combined with *Theorem 3.13* completes the proof. $\square$

Remark 3.19. Let us denote

$$AP(\mathbb{R}_{\mathcal{F}}) = \{f : \mathbb{R} \to \mathbb{R}_{\mathcal{F}}; \ f \text{ is B-almost periodic}\},$$

and for $f \in AP(\mathbb{R}_{\mathcal{F}})$, let us define $\|f\| = \sup\{\|f(t)\|_{\mathcal{F}}; \ t \in \mathbb{R}\}$. By the proof of *Theorem 3.10* we get $\|f\| < +\infty$. Also by *Theorem 3.4* and *Theorems 3.12, 3.17*, $(AP(\mathbb{R}_{\mathcal{F}}), \oplus, \odot)$ is not a linear space, and consequently $(AP(\mathbb{R}_{\mathcal{F}}), \|\cdot\|_{\mathcal{F}})$ is not a normed space. However, endowed with $D^* : AP(\mathbb{R}_{\mathcal{F}}) \times AP(\mathbb{R}_{\mathcal{F}}) \to \mathbb{R}_+$, where

$$D^*(f, g) = \sup_{t \in \mathbb{R}} D(f(t), g(t)),$$

$AP(\mathbb{R}_{\mathcal{F}})$ becomes a complete metric space. Indeed denoting

$$C_b(\mathbb{R}_{\mathcal{F}}) = \{f : \mathbb{R} \to \mathbb{R}_{\mathcal{F}}; f \text{ is continuous and bounded on } \mathbb{R}\},$$

by standard reasonings (taking into account that $(\mathbb{R}_{\mathcal{F}}, D)$ is a complete metric space) it follows that $(C_b(\mathbb{R}_{\mathcal{F}}), D^*)$ is a complete metric space. Then, *Theorems 3.10 and 3.13* show that $AP(\mathbb{R}_{\mathcal{F}})$ is a closed subset of $C_b(\mathbb{R}_{\mathcal{F}})$, i.e. $(AP(\mathbb{R}_{\mathcal{F}}), D^*)$ is a complete metric space.

By similar reasonings with those in the proofs of *Theorems 6.9 and 6.10* in [23, p. 158-160], where we define on $\mathbb{R}_{\mathcal{F}}^m$ the metric

$$D_m(x,y) = \sum_{i=1}^{m} D(x_i, y_i)$$

for all $x = (x_1, ..., x_m)$, and $y = (y_1, ..., y_m) \in \mathbb{R}_{\mathcal{F}}^m$, we can state the following compactness criterion.

Theorem 3.20. *The necessary and sufficient condition that a family $\mathcal{A} \subset AP(\mathbb{R}_{\mathcal{F}})$ be relatively compact is that the following properties hold true:*

(i) $\mathcal{A}$ is equi-continuous;

(ii) $\mathcal{A}$ is equi-almost periodic;

(iii) for any $t \in \mathbb{R}$, the set of values of functions from $\mathcal{A}$ be relatively compact in $\mathbb{R}_{\mathcal{F}}$.

3.3 Harmonics of Almost Periodic Functions in Fuzzy Setting

We start with the concept of integrals of fuzzy functions compatible with the operations introduced in *Section 3.1*

Definition 3.21. *(see [20])* A function $f : [a, b] \to \mathbb{R}_{\mathcal{F}}$, $[a, b] \subset \mathbb{R}$ is said to be Riemann integrable on $[a, b]$, if there exists $I \in \mathbb{R}_{\mathcal{F}}$, with the property: $\forall \varepsilon > 0$, $\exists \delta > 0$, such that for any partition of $[a, b]$, $d : a = x_0 < ... < x_n = b$ of mesh $\nu(d) < \delta$, and for any points $\xi_i \in [x_i, x_{i+1}]$, $0 \le i \le n - 1$, we have

$$D\left(\sum_{i=0}^{n-1}{}^{*} f(\xi_i) \odot (x_{i+1} - x_i), I\right) < \varepsilon,$$

where $\sum^{*}$ means sum with respect to $\oplus$.

In this case we denote

$$I = \int_a^b f(x)\, dx.$$

In order to introduce Fourier series attached to a given function $f \in AP(\mathbb{R}_{\mathcal{F}})$, we need the concept of mean value of f, as follows

Theorem 3.22. *For any $f \in AP(\mathbb{R}_{\mathcal{F}})$, there exists the mean value*

$$M(f) = \lim_{T \to +\infty} \frac{1}{T} \odot \int_0^T f(t)\, dt \in \mathbb{R}_{\mathcal{F}},$$

where the limit is considered in the metric space $(\mathbb{R}_{\mathcal{F}}, D)$, i.e.

$$\exists M(f) \in \mathbb{R}_{\mathcal{F}} \quad with \quad \lim_{T \to +\infty} D\left(M(f), \frac{1}{T} \odot \int_0^T f(t)\, dt\right) = 0.$$

Proof. We follow the ideas in the proof of *Theorem 6.11* in [23, p. 161]. We get

$$\int_\alpha^{\alpha+T} f(t)\, dt = \int_\alpha^\xi f(t)\, dt \oplus \int_\xi^{\xi+T} f(t)\, dt \oplus \int_{\xi+T}^{\alpha+T} f(t)\, dt,$$

which implies

$$D\left(\frac{1}{T} \odot \int_0^T f(t)\, dt, \frac{1}{T} \odot \int_\alpha^{\alpha+T} f(t)\, dt\right)$$

$$\leq D\left(\frac{1}{T} \odot \int_0^T f(t)\, dt, \frac{1}{T} \odot \int_\xi^{\xi+T} f(t)\, dt\right)$$

$$+D\left(\frac{1}{T} \odot \int_\alpha^\xi f(t)\, dt, \tilde{0}\right)$$

$$+D\left(\frac{1}{T} \odot \int_{\xi+T}^{\alpha+T} f(t)\, dt, \tilde{0}\right).$$

Denote $I_1 = \int_{\xi}^{\xi+T} f(t)\, dt$ and $I_2 = \int_0^T f(t+\xi)\, dt$, where by definition

$$D\left(I_1, \sum_{i=0}^{n-1}{}^{*} f(\xi_i') \odot (x_{i+1} - x_i)\right) < \frac{\varepsilon}{2}, \text{ if } \nu(d') < \delta,$$

$$D\left(I_2, \sum_{i=0}^{n-1}{}^{*} f(\xi_i'' + \xi) \odot (y_{i+1} - y_i)\right) < \frac{\varepsilon}{2}, \text{ if } \nu(d'') < \delta,$$

with $d' : \xi = x_0 < ... < x_n = \xi + T$, $\xi_i' \in [x_i, x_{i+1}]$, $0 \le i \le n - 1$, $d'' : 0 = y_0 < ... < y_n = T$, $\xi_i'' \in [y_i, y_{i+1}]$, $0 \le i \le n - 1$. Let us make the choice $x_i = y_i + \xi$, $\xi_i' = \xi_i'' + \xi$, $0 \le i \le n - 1$.

We get:

$$\sum_1{}^{*} = \sum_{i=0}^{*\,n-1} f(\xi_i') \odot (x_{i+1} - x_i) = \sum_{i=0}^{*\,n-1} f(\xi_i'' + \xi) \odot (y_{i+1} - y_i) = \sum_2{}^{*},$$

and therefore

$$D(I_1, I_2) \le D\left(I_1, \textstyle\sum_1^*\right) + D\left(\textstyle\sum_1^*, \textstyle\sum_2^*\right) + D\left(\textstyle\sum_2^*, I_2\right) < \tfrac{\varepsilon}{2} + \tfrac{\varepsilon}{2} = \varepsilon,$$

for any $\varepsilon > 0$,

that is $I_1 = I_2$. It follows (see the properties of D) that

$$D\left(\frac{1}{T} \odot \int_0^T f(t)\, dt, \frac{1}{T} \odot \int_{\xi}^{\xi+T} f(t)\, dt\right)$$

$$= \frac{1}{T} D\left(\int_0^T f(t)\, dt, \int_0^T f(t+\xi)\, dt\right)$$

$$\le \frac{1}{T} \int_0^T D(f(t), f(t+\xi))\, dt$$

Similarly, by $c \odot \tilde{0} = \tilde{0}$, $\forall c \ge 1$, we get

$$D \left(\frac{1}{T} \odot \int_\alpha^\xi f(t)\, dt, \tilde{0} \right) = \frac{1}{T} D \left(\int_\alpha^\xi f(t)\, dt, \tilde{0} \right)$$

$$\leq \frac{1}{T} \int_\alpha^\xi D \left(f(t), \tilde{0} \right) dt$$

and

$$D \left(\frac{1}{T} \odot \int_{\xi+T}^{\alpha+T} f(t)\, dt, \tilde{0} \right) \leq \frac{1}{T} \int_{\xi+T}^{\alpha+T} D \left(f(t), \tilde{0} \right) dt, \text{ for } T \geq 1.$$

Reasoning exactly as in the proof of Theorem 6.11 in [23], we obtain

$$D \left(\frac{1}{T} \odot \int_0^T f(t)\, dt, \frac{1}{T} \odot \int_\alpha^{\alpha+T} f(t)\, dt \right) < \frac{\varepsilon}{2} + 2\frac{A}{T} \cdot l$$

and

$$D \left(\frac{1}{T} \odot \int_0^T f(t)\, dt, \frac{1}{nT} \odot \int_0^{nT} f(t)\, dt \right) < \frac{\varepsilon}{2} + 2\frac{A}{T} \cdot l, \ T > 1$$

The function $\varphi : [1, +\infty) \to \mathbb{R}_{\mathcal{F}}$, $\varphi(T) = \frac{1}{T} \odot \int_0^T f(t)\, dt$, where $f \in AP(\mathbb{R}_{\mathcal{F}})$, is continuous (in the metric D), as product of two continuous functions. This is a property which can be derived by using the properties of D. Indeed, firstly let $f \in AP(\mathbb{R}_{\mathcal{F}})$ be and define $F(T) = \int_0^T f(t)\, dt$, $T \in [1, +\infty)$. Let $T_n \searrow T$, when $n \to +\infty$. We get

$$D \left(F(T_n), F(T) \right) =$$

$$D \left(\int_0^T f(t)\, dt \oplus \int_T^{T_n} f(t)\, dt, \int_0^T f(t)\, dt \oplus \tilde{0} \right)$$

$$\leq D \left(\int_T^{T_n} f(t)\, dt, \tilde{0} \right) \leq \int_T^{T_n} D \left(f(t), \tilde{0} \right) dt \leq \|f\| (T_n - T).$$

If $T_n \nearrow T$, similarly we get $D\left(F\left(T_n\right), F\left(T\right)\right) \leq \|f\|\left(T - T_n\right)$, which proves the continuity of F.

Then, for $T_n \to T$, $T_n, T \in [1, +\infty)$, we get

$$D\left(\frac{1}{T_n} \odot F\left(T_n\right), \frac{1}{T} \odot F\left(T\right)\right) \leq D\left(\frac{1}{T_n} \odot F\left(T_n\right), \frac{1}{T_n} \odot F\left(T\right)\right)$$
$$+ D\left(\frac{1}{T_n} \odot F\left(T\right), \frac{1}{T} \odot F\left(T\right)\right)$$
$$\leq \frac{1}{T_n} D\left(F\left(T_n\right), F\left(T\right)\right)$$
$$+ \left|\frac{1}{T_n} - \frac{1}{T}\right| \cdot D\left(F\left(T\right), \tilde{0}\right) \to 0,$$

when $n \to +\infty$. As a conclusion, φ is continuous on $[1, +\infty)$.

As in the proof of *Theorem 6.11* in [23], take $T_1, T_2 \in [1, +\infty)$ such that $m_1 T_1 = m_2 T_2$, where m_1 and m_2 are two real numbers.

The properties of D allow us to arrive at the inequality (as in the above mentioned proof)

$$D\left(\frac{1}{T_1} \odot \int_0^{T_1} f\left(t\right) dt, \frac{1}{T_2} \odot \int_0^{T_2} f\left(t\right) dt\right) < \varepsilon + 2A\left(\frac{1}{T_1} + \frac{1}{T_2}\right) \cdot l.$$

Continuing the reasoning in [23], we finally arrive at

$$D\left(\frac{1}{T_1} \odot \int_0^{T_1} f\left(t\right) dt, \frac{1}{T_2} \odot \int_0^{T_2} f\left(t\right) dt\right) < 2\varepsilon,$$

for T_1, T_2 sufficiently large $\left(T_1, T_2 > 4Al/\varepsilon\right)$, which proves the theorem. □

Remark 3.23. We can also show that for $f \in AP\left(\mathbb{R}_{\mathcal{F}}\right)$, we have

$$\lim_{T \to +\infty} D\left(M\left(f\right), \frac{1}{T} \odot \int_a^{a+T} f\left(t\right) dt\right) = 0, \quad \text{for all } a \in \mathbb{R}.$$

In what follows we will attach to any function $f \in AP(\mathbb{R}_{\mathcal{F}})$ a Fourier series. We start with the following (see for instance [20]).

Theorem 3.24. $\mathbb{R}_{\mathcal{F}}$ *can be embedded in* $\mathbb{B} = \bar{C}[0,1] \times \bar{C}[0,1]$, *where* $\bar{C}[0,1]$ *is the class of all real valued bounded functions* $f :$ $[0,1] \to \mathbb{R}$ *such that* f *is left continuous for any* $x \in (0,1]$, f *has right limit for any* $x \in [0,1)$ *and* f *is right continuous at* 0. *With the norm* $\|\cdot\| = \sup_{x \in [0,1]} |f(x)|$, $\bar{C}[0,1]$ *is a Banach space.*

Denote $\|\cdot\|_{\mathbb{B}}$ *the usual product norm i.e.*

$$\|(f,g)\|_{\mathcal{B}} = \max \{\|f\|, \|g\|\}.$$

Also denote the embedding by $j : \mathbb{R}_{\mathcal{F}} \to \mathbb{B}$, $j(u) = (u_-, u_+)$. *Then* $j(\mathbb{R}_{\mathcal{F}})$ *is a closed convex cone in* $\mathbb{B}$ *and* j *satisfies the following properties:*

(i) $j(s \odot u \oplus t \odot v) = s \cdot j(u) + t \cdot j(v)$ *for all* $u, v \in \mathbb{R}_{\mathcal{F}}$ *and* $s, t \geq 0$ *(here "·" and "+" denote the scalar multiplication and addition in* $\mathbb{B}$*);*

(ii) $D(u,v) = \|j(u) - j(v)\|_{\mathbb{B}}$ *(i.e.* j *embeds* $\mathbb{R}_{\mathcal{F}}$ *in* $\mathbb{B}$ *isometrically).*

Remark 3.25. Let us denote $\bar{C}_{\mathbb{C}}[0,1] = \{F : [0,1] \to \mathbb{C}; F = F_1 + iF_2, F_1, F_2 \in \bar{C}[0,1]\}$ and $\mathbb{B}_{\mathbb{C}} = \bar{C}_{\mathbb{C}}[0,1] \times \bar{C}_{\mathbb{C}}[0,1]$, where $\mathbb{C}$ represents the set of complex numbers.

It is obvious that $\bar{C}[0,1] \subset \bar{C}_{\mathbb{C}}[0,1]$, $\bar{C}_{\mathbb{C}}[0,1]$ is a complex Banach space endowed with the norm $\|f\| = \sup \{|f(x)| ; x \in [0,1]\}$ and $\mathbb{B}_{\mathbb{C}}$ is a complex banach space endowed with the norm

$$\|F\|_{\mathbb{B}_{\mathbb{C}}} = \max \{\|u\|, \|v\|\}, \forall F = (u,v) \in \mathbb{B}_{\mathbb{C}} = \bar{C}_{\mathbb{C}}[0,1] \times \bar{C}_{\mathbb{C}}[0,1].$$

It is easy to see that $\mathbb{B} = \bar{C}[0,1] \times \bar{C}[0,1] \subset \mathbb{B}_{\mathbb{C}}$ can be isometrically embedded into the complex Banach space $\mathbb{B}_{\mathbb{C}} \times \mathbb{B}_{\mathbb{C}}$ endowed with the product norm $\|(F,G)\|_{\mathbb{B}_{\mathbb{C}} \times \mathbb{B}_{\mathbb{C}}} = \max\{\|F\|_{\mathbb{B}_{\mathbb{C}}}, \|G\|_{\mathbb{B}_{\mathbb{C}}}\}$, where the isometry is defined by

$$I : \mathbb{B} \to \mathbb{B}_{\mathbb{C}} \times \mathbb{B}_{\mathbb{C}}, \; I[(f,g)] = [(f,g),(0,0)],$$

with 0 representing the identical zero function.

Now for the proof of approximation result, we need the following two auxiliary lemmas.

Lemma 3.26. *Let* $f : \mathbb{R} \to \mathbb{R}_{\mathcal{F}}$ *be a B-almost periodic function and* T_m *be a positive valued trigonometric polynomial. Then* $T_m \odot f : \mathbb{R} \to \mathbb{R}_{\mathcal{F}}$ *defined by*

$$(T_m \odot f)(t) = T_m(t) \odot f(t) \text{ for all } t \in \mathbb{R},$$

is B-almost periodic.

Proof. By *Theorem 3.16*, it is enough to prove that $T_m \odot f$ is normal.

Since f is normal, for any sequence of translates $\{f(t+h_n)\}_{n\in\mathbb{N}}$ we have a uniformly convergent subsequence, which we denote $\{f(t+h_{k_n})\}_{n\in\mathbb{N}}$.

Since T_m is a real valued trigonometric polynomial, it is also normal. It follows that $\{T_m(t+h_{k_n})\}_{n\in\mathbb{N}}$ has a convergent subsequence, denoted $\{f(t+h_{l_n})\}_{n\in\mathbb{N}}$.

Then for $n, p \in \mathbb{N}$ we have:

$$D\left((T_m \odot f)(t+h_{l_n}), (T_m \odot f)(t+h_{l_{n+p}})\right)$$

$$D\left(T_m(t+h_{l_n}) \odot f(t+h_{l_n}), T_m(t+h_{l_n}) \odot f(t+h_{l_{n+p}})\right)$$

$$D\left(T_m\left(t+h_{l_n}\right)\odot f\left(t+h_{l_{n+p}}\right), T_m\left(t+h_{l_{n+p}}\right)\odot f\left(t+h_{l_{n+p}}\right)\right).$$

But it is known (see *Lemma 2.2* in [40]) that

$$D\left(a\odot x, b\odot x\right)=|b-a|\cdot\|x\|_{\mathcal{F}},$$

for any $a, b \in \mathbb{R}$ of the same sign and $x \in \mathbb{R}_{\mathbb{F}}$. Then we obtain

$$D\left(\left(T_m\odot f\right)\left(t+h_{l_n}\right), \left(T_m\odot f\right)\left(t+h_{l_{n+p}}\right)\right)\leq$$

$$\left|T_m\left(t+h_{l_n}\right)\right|\cdot D\left(f\left(t+h_{l_n}\right), f\left(t+h_{l_{n+p}}\right)\right)+$$

$$+\left|T_m\left(t+h_{l_n}\right)-T_m\left(t+h_{l_{n+p}}\right)\right|\cdot\left\|f\left(t+h_{l_{n+p}}\right)\right\|_{\mathcal{F}},$$

which proves the lemma. $\square$

Let us now define $p_i : \bar{C}[0,1]\times\bar{C}[0,1]\to\bar{C}[0,1]$, by $p_i((f_1, f_2)) = f_i$, $i=1,2$, for all $(f_1, f_2)\in\bar{C}[0,1]\times\bar{C}[0,1]$. In what follows, the following lemma will be helpful.

Lemma 3.27. *Let $f : \mathbb{R}\to\mathbb{R}_{\mathcal{F}}$. Then f is B-almost periodic (in the sense of Definition 3.8 if and only if $j\circ f : \mathbb{R}\to\bar{C}[0,1]\times\bar{C}[0,1]$ is almost periodic in Bochner's sense (see [23], or [80]), if and only if $p_i\circ j\circ f : \mathbb{R}\to\bar{C}[0,1]$, $i=1,2$ are almost periodic in Bochner's sense.*

Proof. Let us suppose f is almost periodic in the sense of *Definition 3.8 i)*. It follows: $\forall\varepsilon>0$, $\exists l(\varepsilon)>0$ such that any interval of length $l(\varepsilon)$ contains (at least) one point ξ with $D(f(t+\xi), f(t))<\varepsilon$, $\forall t\in\mathbb{R}$.

Because

$$D\left(f(t+\tau), f(t)\right) = \|(j \circ f)(t+\xi) - (j \circ f)(t)\|_{\mathbb{B}},$$

we obtain the first equivalence.

To prove the second equivalence, let us first assume that $j \circ f$ is almost periodic in Bochner's sense. Because

$$\|(j \circ f)(t+\tau) - (j \circ f)(t)\|_{\mathbb{B}} = \max\{\|(p_1 \circ j \circ f)(t+\xi)$$
$$-(p_1 \circ j \circ f)(t)\|,$$
$$\|(p_2 \circ j \circ f)(t+\xi)$$
$$-(p_2 \circ j \circ f)(t)\|\},$$

it easily follows that $p_i \circ j \circ f : \mathbb{R} \to \bar{C}[0,1]$, $i = 1,2$, are also almost periodic in Bochner's sense.

The converse implication is immediate by the above relation, which proves the lemma. $\square$

The next result called *approximation property*, is in fact the converse of *Theorem 3.18*, and represents one of the most important property of $f \in AP(\mathbb{R}_{\mathcal{F}})$.

Theorem 3.28. *Let $f : \mathbb{R} \to \mathbb{R}_{\mathcal{F}}$ a B-almost periodic function. Then there exists a sequence of generalized trigonometric polynomials T_m such that $T_m \to f$ uniformly on $\mathbb{R}$.*

Proof. Let $j \circ f : \mathbb{R} \to \mathbb{B}$ be the embedding defined in *Theorem 3.24*. By *Theorem 6.13* [23], the mean values

$$a_1(\lambda) = M\{\cos \lambda t \cdot j \circ f(t)\}$$

and

$$a_2(\lambda) = M\{\sin \lambda t \cdot j \circ f(t)\}$$

exist. Indeed let $j_1 : \mathbb{R}_{\mathcal{F}} \to \mathbb{B}_{\mathbb{C}} \times \mathbb{B}_{\mathbb{C}}$, where $j_1 = I \circ j$ with I the natural embedding $I : \mathbb{B} \to \mathbb{B}_{\mathbb{C}} \times \mathbb{B}_{\mathbb{C}}$.

For $j_1 \circ f$ we apply again *Theorem 6.13* [23] and we obtain that $a(\lambda) = M\{j_1 \circ f(t) \cdot e^{-i\lambda t}\} \neq 0$ only for a set at most countable of complex numbers $\lambda_1, \lambda_2, \ldots$

In what follows we use the ideas in the proof of *Theorem 6.15* [23].

Let us consider the Féjer kernel $K_n(t) = \frac{1}{n} \cdot \frac{\sin^2 \frac{nt}{2}}{\sin^2 \frac{t}{2}}$. It is well known that $K_n(t) = \sum_{\nu=-n}^{n} \left(1 - \frac{|\nu|}{n}\right) \cos \nu t$ is an even trigonometric polynomial. Let $\beta_1, \ldots, \beta_n, \ldots$ be a basis of the Fourier exponents $\lambda_1, \ldots, \lambda_k, \ldots$ (i.e. there exist $r_1, \ldots, r_n \in \mathbb{Q}$ such that $\lambda_k = r_1\beta_1 + \ldots + r_n\beta_n$). Let $\mathcal{K}_m(t) = K_{m!^2}\left(\frac{\beta_1 t}{m!}\right) \ldots K_{m!^2}\left(\frac{\beta_m t}{m!}\right) \geq 0$.

Observe that $\mathcal{K}_m(t)$ is also an even trigonometric polynomial. Let $\mathcal{K}_m \cdot (I \circ j \circ f) : \mathbb{R} \to \mathbb{B}_{\mathbb{C}} \times \mathbb{B}_{\mathbb{C}}$. By *Lemma 3.27* we obtain $j \circ f$ is almost periodic and it is easy to see that $I \circ j \circ f$ is almost periodic too.

By the proof of *Theorem 6.15* [23] we obtain

$$\sigma_m(t) = \lim_{T \to \infty} \frac{1}{T} \int_0^T \mathcal{K}_m(u)(I \circ j \circ f)(u+t)du$$

converges uniformly to $I \circ j \circ f$.

By the form of the embedding I, it follows that $\sigma'_m(t) = \lim_{T \to \infty} \frac{1}{T} \int_0^T \mathcal{K}_m(u) \cdot (j \circ f)(t+u) du$ converges uniformly to $j \circ f$.

Let us consider $\sigma_m^{\mathcal{F}}(t) = M\{\mathcal{K}_m(u) \odot f(t+u)\}$ which exists by *Theorem 3.12* and *Lemma 3.26*. Then we obtain:

$$D\left(\sigma_m^{\mathcal{F}}(t), f(t)\right) = \left\|(j \circ \sigma_m^{\mathcal{F}})(t) - (j \circ f)(t)\right\|_{\mathbb{B}} \leq$$

$$\left\|(j \circ \sigma_m^{\mathcal{F}})(t) - \sigma_m'(t)\right\|_{\mathbb{B}} + \left\|\sigma_m'(t) - (j \circ f)(t)\right\|_{\mathbb{B}}.$$

It follows that

$$\left\|(j \circ \sigma_m^{\mathcal{F}})(t) - \sigma_m'(t)\right\|_{\mathbb{B}} =$$

$$\left\| j\left(\lim_{T \to \infty} \frac{1}{T} \int_0^T \mathcal{K}_m(u) \odot f(u+t)\, du \right) - \right.$$

$$\left. - \lim_{T \to \infty} \frac{1}{T} \int_0^T \mathcal{K}_m(u) \cdot (j \circ f)(u+t)\, du \right\|_{\mathbb{B}}.$$

It is easy to see that

$$j\left(\lim_{T \to \infty} \frac{1}{T} \int_0^T \mathcal{K}_m(u) \odot f(u+t)\, du \right) =$$

$$= \lim_{T \to \infty} \frac{1}{T} j\left(\int_0^T \mathcal{K}_m(u) \odot f(u+t)\, du \right).$$

By *Theorem 3.2* [20] and since $\mathcal{K}_m$ is positive it follows:

$$(j \circ \sigma_m^{\mathcal{F}})(t) = \lim_{T \to \infty} \frac{1}{T} \int_0^T j\left(\mathcal{K}_m(u) \odot f(u+t)\right) du =$$

$$= \lim_{T \to \infty} \frac{1}{T} \int_0^T \mathcal{K}_m(u) \cdot (j \circ f)(u+t)\, du = \sigma_m'(t).$$

This leads us to $D\left(\sigma_m^{\mathcal{F}}(t), f(t)\right) < \frac{\varepsilon}{2}$ for sufficiently large m and any $t \in \mathbb{R}$. Observe that $\sigma_m^{\mathcal{F}}(t)$ is not necessarily a fuzzy trigonometric polynomial, since by *Theorem 3.4*, (iii), for general coefficients, the integral does not necessarily commute with the sum related to "$\oplus$".

However $\sigma_m^{\mathcal{F}}(t)$ is a periodic function (since $\sigma_m'(t)$ is periodic). By *Theorem 3.1* [3] it follows that there exists a generalized fuzzy trigonometric polynomial P_m such that

$$D\left(\sigma_m^{\mathcal{F}}(t), P_m(t)\right) < \frac{\varepsilon}{2}.$$

Finally we obtain

$$D\left(P_m(t), f(t)\right) \leq D\left(P_m(t), \sigma_m^{\mathcal{F}}(t)\right) + D\left(\sigma_m^{\mathcal{F}}(t), f(t)\right) < \varepsilon.$$

Which completes the proof. $\square$

In conclusion we infer that , *Theorems 3.16, 3.18 and 3.28* lead to the equivalence of all three concepts in *Definition 3.8*, in other words we have:

Corollary 3.29. *The set of all B-almost periodic fuzzy-number-valued functions coincides with the set of all continuous normal functions and with the set of all functions with approximation property.*

3.4 Applications to Fuzzy Differential Equations

We start this section by the illustration of the idea of propagation of almost periodicity from the fuzzy input data to the solutions of a fuzzy differential equation as follows:

Theorem 3.30. *Let $f \in AP(\mathbb{R})$ that is f is (Bochner) almost periodic assume $c \in \mathbb{R}_{\mathcal{F}}$ is a fuzzy real number. If $f(t) \geq 0$, for all $t \in \mathbb{R}$, then the function $y : \mathbb{R} \to \mathbb{R}_{\mathcal{F}}$ given by*

$$y(t) = c \odot \int_{-\infty}^{t} e^{t-u} f(u) du, \ t \in \mathbb{R} \qquad (3.1)$$

is B-almost periodic and satisfies the fuzzy differential equation

$$y'(t) \oplus y(t) = c \odot f(t),$$

for all $t \in \Omega$, where $\Omega = \left\{ t \in \mathbb{R}; f(t) > \int_{-\infty}^{t} e^{u-t} f(u) du \right\}$.

Here $y'(t)$ is defined as the common value of the following limits in the metric D supposed that exist, together with the H-differences $y(t+h) - y(t)$, $y(t) - y(t-h)$, $h > 0$ (recall that the H-difference $y(t+h) - y(t)$ exists, if exists $a \in \mathbb{R}_{\mathcal{F}}$ such that $y(t+h) = y(t) \oplus a$),

$$\lim_{h \searrow 0} \frac{y(t+h) - y(t)}{h} = \lim_{h \searrow 0} \frac{y(t) - y(t-h)}{h} = h'(t)$$

(see e.g. Definition 3.3 [20]).

Proof. Let us denote $F(t) = \int_{-\infty}^{t} e^{u-t} f(u) du$, $t \in \mathbb{R}$. Then clearly F is (Bochner) almost periodic with the same ε-period than the function f. Now since

$$D(c \odot F(t), c \odot F(t+\tau)) \leq |F(t) - F(t+\xi)| \cdot D(c, \widetilde{0}),$$

where $\widetilde{0} \in \mathbb{R}_{\mathcal{F}}$, it is immediate that $y(t) = c \odot F(t)$ is B-almost periodic (i.e. in the sense of *Definition 3.8, (i)*).

It remains to prove that $y(t)$ is differentiable and satisfies the fuzzy differential equation.

Indeed, let $h > 0$. By hypothesis we have $F'(t+h), F(t) \geq 0$, for all $t \in \mathbb{R}$.

On the other hand

$$F'(t) = f(t) - \int_{-\infty}^{t} e^{u-t} f(u) du > 0, \forall t \in \Omega,$$

which implies

$$F(t + h) - F(t) = \int_{-\infty}^{t+h} e^{u-(t+h)} f(u) du - \int_{-\infty}^{t} e^{u-t} f(u) du$$

$$= H(t, h) > 0,$$

for $t \in \Omega$ and $h > 0$, sufficiently small.

By *Theorem 3.4*, (iii), we get

$$c \odot F(t + h) = c \odot F(t) \oplus c \odot H(t, h),$$

that is

$$c \odot F(t + h) - c \odot F(t) = y(t + h) - y(t) = c \odot H(t, h).$$

Multiplying by $\frac{1}{h}$, in view of *Theorem 3.4*, (v), gives

$$\frac{y(t + h) - y(t)}{h} = c \odot \frac{H(t, h)}{h}.$$

Passing to the limit as $h \searrow 0$ in the metric space $(\mathbb{R}, D)$, we easily obtain that

$$\lim_{h \searrow 0} \frac{y(t + h) - y(t)}{h} = c \odot \left[f(t) - \int_{-\infty}^{t} e^{u-t} f(u) du \right].$$

Similarly we obtain

$$\lim_{h \searrow 0} \frac{y(t) - y(t - h)}{h} = c \odot \left[f(t) - \int_{-\infty}^{t} e^{u-t} f(u) du \right],$$

that is

$$y'(t) = c \odot \left[f(t) - \int_{-\infty}^{t} e^{u-t} f(u) du \right].$$

Then, again in view of *Theorem 3.4*, (iii), we get

$$y(t) \oplus y'(t) = c \odot \int_{-\infty}^{t} e^{u-t} f(u) du \oplus c \odot \left[f(t) - \int_{-\infty}^{t} e^{u-t} f(u) du \right]$$

$$= c \odot \left[\int_{-\infty}^{t} e^{u-t} f(u) du + f(t) - \int_{-\infty}^{t} e^{u-t} f(u) du \right] = c \odot f(t),$$

for all $t \in \Omega$, which proves the theorem. $\square$

Example 3.31. A simple example satisfying the above *Theorem 3.30* is $f(t) = 3 + \cos t + \cos(t\sqrt{2}) > 0, \forall t \in \mathbb{R}$, which is almost periodic (but it is not periodic) on $\mathbb{R}$.

Then according to *Theorem 3.4*, we get

$y(t) = c \odot F(t)$, $y : \mathbb{R} \to \mathbb{R}_{\mathcal{F}}$, with $c \in \mathbb{R}_{\mathcal{F}}$ and $F(t) = \int_{-\infty}^{t} e^{u-t} \left[3 + \cos u + \cos(u\sqrt{2}) \right] du$, is B-almost periodic. On the other hand, we have

$$F(t) = \int_{-\infty}^{t} e^{u-t} \left[3 + \cos u + \cos(u\sqrt{2}) \right] du =$$

$$= e^{-t} \left\{ 3e^t + \frac{e^t (\cos t + \sin t)}{2} + \frac{e^t [\cos(t\sqrt{2}) + \sqrt{2} \sin(t\sqrt{2})]}{3} \right\} =$$

$$= 3 + \frac{\cos t + \sin t}{2} + \frac{\cos(t\sqrt{2}) + \sqrt{2} \sin(t\sqrt{2})}{3}.$$

The condition $f(t) > \int_{-\infty}^{t} e^{u-t} f(u) du$ becomes

$$3 + \cos t + \cos(t\sqrt{2}) > 3 + \frac{\cos t + \sin t}{2} + \frac{\cos(t\sqrt{2}) + \sqrt{2} \sin(t\sqrt{2})}{3},$$

which is equivalent to

$$3(\cos t - \sin t) + 2\sqrt{2}[\cos(t\sqrt{2}) - \sin(t\sqrt{2})] + (4 - 2\sqrt{2}) \cos(t\sqrt{2}) > 0,$$

Simple considerations prove that for $t \in \left(-\frac{\pi}{2\sqrt{2}}, \frac{\pi}{4\sqrt{2}}\right)$, the above inequality holds, that is $\left(-\frac{\pi}{2\sqrt{2}}, \frac{\pi}{4\sqrt{2}}\right) \subset \Omega$.

Denoting $E = \bigcup_{k \in \mathbb{Z}} \left(-\frac{\pi}{2\sqrt{2}} + 2k\pi, \frac{\pi}{4\sqrt{2}} + 2k\pi\right)$, actually we have $E \subset \Omega$ and therefore for all $t \in E$,

$$y(t) = c \odot \left[3 + \frac{\cos t + \sin t}{2} + \frac{\cos(t\sqrt{2}) + \sqrt{2}\sin(t\sqrt{2})}{3}\right],$$

satisfies the fuzzy differential equation

$$y'(t) \oplus y(t) = c \odot [3 + \cos t + \cos(t\sqrt{2})], \quad t \in E. \square$$

3.5 Bibliographical Remarks and Open Problems

In this chapter the main properties of real-valued almost periodic functions were extended to fuzzy-number-valued almost periodic functions as in [23], [80].

Applications to dynamical systems as in *Section 1.6.2* are also possible (see B. Bede and S. G. Gal [40] for details).

It would be interesting to use other concepts of differentiabilty in fuzzy settings in the study of fuzzy differential equations as indicated by B. Bede and S. Gal ([9, 40]).

4

Almost Automorphy in Fuzzy Setting

4.1 Introduction

The purpose of this chapter is to extend the main properties of
Banach-space-valued almost automorphic functions as presented
in [80], to fuzzy-number-valued almost automorphic functions.
This is done in *Section 4.3* below.

Although majority of proofs follow standard ideas of proofs in
Chapter 2 of [80], however, their adaptation requires a careful ma-
nipulation of the properties in the complete metric spaces $(\mathbb{R}_{\mathcal{F}}, D)$
and $(X, \oplus, \odot, d)$ (see *Section 3.1* for details).

Also the facts that $\mathbb{R}_{\mathcal{F}}$ (and X) with respect to addition $\oplus$ is
not a group and that with respect to real scalars multiplication
$\odot$ too is not a linear space (the distributivity of sum $+$ of reals
with respect to $\odot$ does not hold in general, it holds only if the real
scalars are all ≥ 0 or all ≤ 0), require changes of some concepts
and proofs.

Section 4.2 contains new spaces constructed with the aid of $(\mathbb{R}_{\mathcal{F}}, D)$, with properties similar to those of $(\mathbb{R}_{\mathcal{F}}, D)$, fact which permits to enlarge considerably the applicability of the theory.

In *Section 4.4* we present some applications to fuzzy differential equations.

4.2 Preliminaries

With the aid of $(\mathbb{R}_{\mathcal{F}}, \oplus, \odot, D)$ introduced in *Chapter 3*, let us define new spaces, as follows.

(1) $C([a, b]; \mathbb{R}_{\mathcal{F}})$-the space of all continuous functions $f : [a, b] \to \mathbb{R}_{\mathcal{F}}$, endowed with the metric $D^*(f, g) = \sup\{D(f(x), g(x)); x \in [a, b]\}$ (and the natural operations induced by those in $\mathbb{R}_{\mathcal{F}}$;)

(2) For $1 \leq p < +\infty$, $L^p([a, b]; \mathbb{R}_{\mathcal{F}})$ the space of strongly measurable functions on $[a, b]$, $f : [a, b] \to \mathbb{R}_{\mathcal{F}}$, such that

$$(L) \quad \int_a^b D^p(\tilde{0}, f(x)) dx < +\infty,$$

endowed with the metric

$$D_p(f, g) = \left((L) \int_a^b D^p(f(x), g(x)) dx \right)^{1/p};$$

(3) For $1 \leq p < +\infty$,

$$l_{\mathbb{R}_{\mathcal{F}}}^p = \{x = (x_n); x_n \in \mathbb{R}_{\mathcal{F}}, \forall n \in \mathbb{N}, \sum_{n=1}^{\infty} ||x_n||_{\mathcal{F}}^p < +\infty\},$$

endowed with the metric

$$\rho(x, y) = \left(\sum_{n=1}^{\infty} D^p(x_n, y_n) \right)^{1/p}, \forall x = (x_n), y = (y_n) \in l_{\mathbb{R}_{\mathcal{F}}}^p;$$

(4) $m_{\mathbb{R}_{\mathcal{F}}}$-the space of all sequences of fuzzy numbers $x = (x_n)_n$, bounded in the "norm" $||.||_{\mathcal{F}}$, i.e. there exists $M > 0$ (depending on x) such that $||x_n||_{\mathcal{F}} \leq M$, for all $n \in \mathbb{N}$, endowed with the metric $\mu(x,y) = \sup\{D(x_n, y_n); n \in \mathbb{N}\}$, for all $x = (x_n)_n, y = (y_n)_n \in \mathbb{R}_{\mathcal{F}}$;

(5) $c_{\mathbb{R}_{\mathcal{F}}}$-the space of all convergent sequences (in the metric D) of fuzzy numbers and $c_{\mathbb{R}_{\mathcal{F}}}^{\tilde{0}}$-the space of all convergent to $\tilde{0}$ sequences of fuzzy numbers, both endowed with the metric μ from the above case;

(6) First we need the following known definition :

$f : [a, b] \to \mathbb{R}_{\mathcal{F}}$ is called Hukuhara differentiable on $x \in (a, b)$, if there is $\delta > 0$ such that for all $0 < h \leq \delta$ there exist the quantities $f(x+h) \ominus f(x)$, $f(x) \ominus f(x-h)$ and $l \in \mathbb{R}_{\mathcal{F}}$ denoted by $f'(x)$, such that

$$\lim_{h \searrow 0} D(\frac{1}{h} \odot (f(x+h) \ominus f(x)), f'(x)) =$$
$$\lim_{h \searrow 0} D(\frac{1}{h} \odot (f(x) \ominus f(x-h)), f'(x)) = 0.$$

For $p \in \mathbb{N}$, one considers the space

$$C^p([a,b]; \mathbb{R}_{\mathcal{F}}) = \{f : [a,b] \to \mathbb{R}_{\mathcal{F}}; \exists f^{(p)} \in C([a,b]; \mathbb{R}_{\mathcal{F}})\},$$

endowed with the metric $D_p^*(f,g) = \sum_{i=1}^p D^*(f^{(i)}, g^{(i)})$, where the derivative is in Hukuhara sense.

The class of Hukuhara differentiable fuzzy-number-valued functions can considerably be enlarged, with the aid of the following more general definition of differentiability introduced in [40] :

A function $f : (a, b) \to \mathbb{R}_{\mathcal{F}}$ is called generalized differentiable on $t \in (a, b)$ if:

(i) There exist $f(t + h) \ominus f(t)$, $f(t) \ominus f(t - h)$, for all $h > 0$ sufficiently small and there exist

$$\lim_{h \searrow 0} \frac{f(t + h) \ominus f(t)}{h} = \lim_{h \searrow 0} \frac{f(t) \ominus f(t - h)}{h} = f'(t) \in \mathbb{R}_{\mathcal{F}}$$

or

(ii) There exist $f(t) \ominus f(t + h)$, $f(t - h) \ominus f(t)$, for all $h > 0$ sufficiently small and there exist

$$\lim_{h \searrow 0} \frac{f(t) \ominus f(t + h)}{-h} = \lim_{h \searrow 0} \frac{f(t - h) \ominus f(t)}{-h} = f'(t) \in \mathbb{R}_{\mathcal{F}}$$

or

(iii) There exist $f(t + h) \ominus f(t)$, $f(t - h) \ominus f(t)$, for all $h > 0$ sufficiently small and there exist

$$\lim_{h \searrow 0} \frac{f(t + h) \ominus f(t)}{h} = \lim_{h \searrow 0} \frac{f(t - h) \ominus f(t)}{-h} = f'(t) \in \mathbb{R}_{\mathcal{F}}$$

or

(iv) There exist $f(t) \ominus f(t - h)$, $f(t) \ominus f(t + h)$, for all $h > 0$ sufficiently small and there exist

$$\lim_{h \searrow 0} \frac{f(t) \ominus f(t - h)}{h} = \lim_{h \searrow 0} \frac{f(t) \ominus f(t + h)}{-h} = f'(t) \in \mathbb{R}_{\mathcal{F}}$$

(Here all the limits are considered in the metric D and h or $-h$ at denominators, in fact means $\frac{1}{h} \odot$ or $-\frac{1}{h} \odot$, respectively).

It is evident that Hukuhara differentiability implies the generalized differentiability but the converse implication does not hold. Also, the space

$$C^p([a, b]; \mathbb{R}_{\mathcal{F}}) = \{f : [a, b] \to \mathbb{R}_{\mathcal{F}}; \exists f^{(p)} \in C([a, b]; \mathbb{R}_{\mathcal{F}})\},$$

can be considered for the generalized differentiability too.

Remark 4.1. All these spaces have been studied in [38], where as a conclusion it is derived that if we denote by $(X, \oplus, \odot, d)$ any from the spaces considered by the previous points 1)-6), endowed of course with the natural operations $\oplus, \odot$ induced by those $\oplus$ and $\odot$ in $\mathbb{R}_{\mathcal{F}}$, then it has all the properties of $(\mathbb{R}_{\mathcal{F}}, \oplus, \odot, D)$, presented in *Chapter 3.*

Also, any finite Cartesian product of the spaces considered above (including $(\mathbb{R}_{\mathcal{F}}, D)$ too) endowed with the "box metric" (i.e. $d = \max\{\rho_i; i\}$) and with the natural induced operations $\oplus$ and $\odot$, has all the above mentioned properties of $(\mathbb{R}_{\mathcal{F}}, \oplus, \odot, D)$.

Finally, let us note that the definitions of Hukuhara differentiability and of generalized differentiability, can similarly be considered if the function $f : (a, b) \to \mathbb{R}_{\mathcal{F}}$ is replaced by $f : (a, b) \to X$, where $(X, \oplus, \odot, d)$ is any from the above mentioned spaces.

Let us recall now some elements of operator theory and semigroup of operators on $(X, \oplus, \odot, d)$ in [38], where $(X, \oplus, \odot, d)$ denotes any from the above mentioned spaces (including the case $X = \mathbb{R}_{\mathcal{F}}$).

Definition 4.2. *(i) $A : X \to X$ is called linear operator if*

$$A(\lambda \odot x \oplus \mu \odot y) = \lambda \odot A(x) \oplus \mu \odot A(y),$$

for all $\lambda, \mu \in \mathbb{R}$ and all $x, y \in X$.

(ii) The family $T = \{T(t)), t \in \mathbb{R}_+\}$ of continuous linear operators on X is called C_0-semigroup if :

1) For all $x \in X$, the mapping $T(t)(x) : \mathbb{R}_+ \to X$ is continuous with respect to $t \geq 0$;

2) $T(t+s) = T(t)[T(s)]$, for all $t, s \in \mathbb{R}_+$;

3) $T(0) = I$, where I is the identity operator on X ;

(iii) If $A : X \to X$ is a linear operator, then it is called generator of the C_0-semigroup, if for all $x \in X$, there exists $T(t)(x) \ominus x$ and $\lim_{h \searrow 0} d(A(x), \frac{1}{t} \odot [T(t)(x) \ominus x]) = 0$.

Theorem 4.3. *([38])*

(i) If $A : X \to X$ is linear and continuous on $\tilde{0}_X$, then for all $x \in X$ we have

$$||A(x)||_{\mathcal{F}} \le |||A|||_{\mathcal{F}}||x||_{\mathcal{F}},$$

where $|||A|||_{\mathcal{F}} = \sup\{||A(x)||_{\mathcal{F}}; x \in X, ||x||_{\mathcal{F}} \le 1\} \in \mathbb{R}$. If A is linear on X and continuous on $\tilde{0}_X$, then it does not follow the continuity of A on the whole space X.

All these considerations remain valid if instead to be linear, A is supposed to be only additive (i.e. $A(x \oplus y) = A(x) \oplus A(y)$) and positive homogeneous (i.e $A(\lambda \odot x) = \lambda \odot A(x)$, for all $\lambda \ge 0$).

(ii) (Uniform boundedness principle) Let $\{A_j, j \in J\}$ be a family of additive, positive homogeneous and continuous operators on X. If $\{A_j, j \in J\}$ is pointwise bounded (i.e. for any $x \in X$, there exists $M_x \in \mathbb{R}$ such that $||A_j(x)||_{\mathcal{F}} \le M_x, \forall j \in J$), then there exists a real number $M > 0$ such that $|||A_j|||_{\mathcal{F}} \le M, \forall j \in J$.

(iii) For any A, linear and continuous operator on X, can be defined the linear and continuous operators $T(t) = e^{t \odot A}, t \in \mathbb{R}$ by

$$\lim_{m \to +\infty} d(T(t), \sum_{p=0}^{m} {}^{*} \frac{t^p}{p!} \odot A^p) = 0,$$

where $\sum^*$ is the sum with respect to $\oplus$ and $A^0 = I, A^p = A^{p-1} \circ A, p = 2, 3, ...,$ satisfying the following properties:

1) The family $T = \{T(t)), t \in \mathbb{R}_+\}$ is C_0-semigroup on X (as in the above Definition 4.2, (ii)) and in addition, $T(t)$ is continuous for $t < 0$ too. Also, the property $T(t + s) = T(t)[T(s)]$ holds for all $t, s < 0$, but does not hold if t and s are of contrary signs.

2) $T(t)$ is generalized differentiable with respect to all $t \in \mathbb{R}$, with the derivative equal to $A[T(t)]$. More exactly, it is Hukuhara differentiable with respect to $t \in \mathbb{R}_+$, i.e.

$$\lim_{h \searrow 0} d(\frac{1}{h} \odot (T(t+h)(x) \ominus T(t)(x)), A[T(t)(x)]) =$$

$$\lim_{h \searrow 0} d(\frac{1}{h} \odot (T(t)(x) \ominus T(t-h)(x)), A[T(t)(x)]) = 0,$$

and generalized differentiable with respect to $t < 0$, i.e.

$$\lim_{h \searrow 0} d(-\frac{1}{h} \odot (T(t)(x) \ominus T(t+h)(x)), A[T(t)(x)]) =$$

$$\lim_{h \searrow 0} d(-\frac{1}{h} \odot (T(t-h)(x) \ominus T(t)(x)), A[T(t)(x)]) = 0,$$

for all $x \in X$.

3) $\lim_{t \searrow 0} d(\frac{1}{t} \odot [T(t)(x) \ominus x], A(x))] = 0$, for all $x \in X$.

4) If $u_0 \in X$ and $g : \mathbb{R} \to X$ is continuous on $\mathbb{R}$, then

$$u(t) = T(t)(u_0) \oplus \int_0^t T(t-s)g(s)ds$$

is generalized differentiable on $\mathbb{R}$ (more exactly it is Hukuhara differentiable on $\mathbb{R}_+$ and generalized differentiable for $t < 0$ as in the above point 2)) and satisfies $u(0) = u_0$, $u'(t) = A[u(t)] \oplus g(t), \forall t \in R$, where $u'(t)$ denotes the generalized derivative.

Here the integral for functions defined on a compact interval with values in X is considered in the Riemann (classical) sense (see for instance [80]).

4.3 Basic Definitions and Properties

Everywhere in the rest of the chapter, $(X, \oplus, \odot, d)$ will denote any from the spaces (including $(\mathbb{R}_\mathcal{F}, \oplus, \odot, D)$) considered by the previous section.

Starting from the Bochner-kind definition for the almost automorphy, in this section we develop a theory of almost automorphic functions with values in $(X, \oplus, \odot, d)$, similar to that for Banach-space valued functions (see [80]).

We rewrite *Definition 1.29* as follows.

Definition 4.4. *. We say that a continuous function $f : \mathbb{R} \to X$, is almost automorphic, if every sequence of real numbers (r_n), contains a subsequence (s_n), such that there exists $g(t) \in X$ with the property*

$$\lim_{n \to +\infty} d(g(t), f(t + s_n)) = \lim_{n \to +\infty} d(g(t - s_n), f(t)) = 0.$$

for each $t \in \mathbb{R}$

Remark 4.5. As in the classical theory, the above convergence on $\mathbb{R}$ is pointwise. The concept of almost automorphy in *Definition 4.4,* is more general than almost periodicity in Bochner's sense. Indeed, if the convergence in *Definition 4.4* is uniform on $\mathbb{R}$, then according to the theory developped in *Chapter 3*, we get the almost periodicity.

Note that although the proof of *Theorem 3.16* in [40] is given for functions with values in $\mathbb{R}_{\mathcal{F}}$, however because of the considerations in the previous *Section 4.2*, it remains valid for functions with values in X.

Also, there exist almost automorphic functions which are not almost periodic. For example, take $X = \mathbb{R}_{\mathcal{F}}$ and define $f(x) = c \odot g(x)$, $x \in \mathbb{R}$, where $c \in \mathbb{R}_{\mathcal{F}}$ and $g : \mathbb{R} \to \mathbb{R}$ is an example in e.g. [89], of almost automorphic function which is not almost periodic. Then it easily follows that f is almost automorphic in the sense of the above *Definition 4.4* but it is not almost periodic in Bochner's sense, i.e. as in *Chapter 3 Definition 3.8*.

The following elementary properties hold.

Theorem 4.6. . *If $f, f_1, f_2 : \mathbb{R} \to X$ are almost automorphic functions then we have :*

(i) $f_1 \oplus f_2$ is almost automorphic ;

(ii) $c \odot f$ is almost automorphic for every scalar $c \in \mathbb{R}$;

(iii) $f_a(t) := f(t + a), \forall t \in \mathbb{R}$ is almost automorphic for each fixed $a \in \mathbb{R}$;

(iv) f is bounded, i.e. $\sup\{||f(t)||_{\mathcal{F}}; t \in \mathbb{R}\} < +\infty$;

(v) The range $R_f = \{f(t); t \in \mathbb{R}\}$ is relatively compact in the complete metric space (X, d);

(vi) The function h defined by $h(t) := f(-t), t \in \mathbb{R}$ is almost automorphic;

(vii) If $f(t) = \tilde{0}_X$ for all $t > a$ for some real number a, then $f(t) = \tilde{0}_X$ for all $t \in \mathbb{R}$;

(viii) If $A : X \to Y$ is continuous, where Y also is any from the spaces considered in Section 4.2, then $A(f) : \mathbb{R} \to Y$ is almost automorphic.

(ix) Assume that $A : X \to X$ is a continuous linear operator on X and $x(t) = e^{t \odot A}(x_0)$, $t \in \mathbb{R}$ is almost automorphic for some $x_0 \in X$. If there exists a bounded subset K of $\mathbb{R}_+$ such that $\inf\{||x(t)||_{\mathcal{F}}; t \in K\} = 0$, then $x(t) = \tilde{0}_X$, for all $t \in \mathbb{R}$;

(x) Let $h_n : \mathbb{R} \to X, n \in \mathbb{N}$ be a sequence of almost automorphic functions such that $h_n(t) \to h(t)$ when $n \to +\infty$, uniformly in $t \in \mathbb{R}$. Then h is almost automorphic.

Proof. (i) It is immediate from the property,

$$d\left(u \oplus v, w \oplus e\right) \leq d\left(u, w\right) + d\left(v, e\right), \ \forall \, u, v, w, e \in X,$$

and from *Definition 4.4.*

(ii) It is immediate from the property,

$$d\left(c \odot u, c \odot v\right) = |c|\, d\left(u, v\right), \ \forall \, u, v \in X, \forall c \in \mathbb{R},$$

and from *Definition 4.4.*

(iii) It is immediate by *Definition 4.4.*

(iv) We follow the lines of proof of *Theorem 1.31 iv).* Indeed, let us suppose that $\sup\{||f(t)||_{\mathcal{F}}; t \in \mathbb{R}\} = +\infty$, i.e. there exists a sequence of real numbers $(r_n)_n$ such that $||f(r_n)||_{\mathcal{F}} \to +\infty$, when $n \to +\infty$.

Since f is almost automorphic, by *Definition 4.4* for $t = 0$, there exists a subsequence (s_n) of (r_n) such that

$$\lim_{n \to +\infty} d(g(0), f(s_n)) = 0,$$

where $g(0) \in X$.

By passing to limit with $n \to +\infty$ in the relations

$$\|f(s_n)\|_{\mathcal{F}} = d(\tilde{0}, f(s_n)) \leq d(\tilde{0}, g(0)) + d(g(0), f(s_n)),$$

we get the contradiction $d(\tilde{0}, g(0)) = +\infty$.

(v) Let $(f(r_n))$ be an arbitrary sequence in X. From *Definition 4.4*, there exists a subsequence (s_n) of (r_n) such that

$$\lim_{n \to +\infty} d(g(0), f(s_n)) = 0,$$

i.e. $(f(s_n))_n$ is a convergent subsequence of $(f(r_n))$ in the complete metric space (X, d), which proves that R_f is relatively compact in (X, d).

(vi) The proof is similar to the proof of *Theorem 2.1.4* in [80].

(vii) The proof is identical to the proof of *Theorem 2.1.8* in [80].

(viii) It is an immediate consequence of *Definition 4.4* and continuity of A.

(ix) We follow some ideas in the proof of *Theorem 2.1.9* [80], but adapted to our case.

Suppose that K is a bounded subset of $\mathbb{R}_+$ such that

$$\inf\{\|x(t)\|_{\mathcal{F}}; t \in K\} = 0.$$

We can find a sequence $s_n \in K, n \in \mathbb{N}$ and $y : \mathbb{R} \to X$, such that $\lim_{n \to +\infty} \|x(s_n)\|_{\mathcal{F}} = 0$ and

$$\lim_{n \to +\infty} d(y(t), x(t + s_n)) = \lim_{n \to +\infty} d(y(t - s_n), x(t)) = 0,$$

pointwise on $\mathbb{R}$.

Because K is bounded, there exists $M > 0$ such that

$$0 \leq s < M, \forall s \in K.$$

From the above *Theorem 4.3*, we get

$$x(t + s_n) = e^{(t+s_n)\odot A}(x_0)$$
$$= e^{t\odot A \oplus s_n \odot A}(x_0)$$
$$= e^{t\odot A}(e^{s_n \odot A}(x_0))$$
$$= e^{t\odot A}(x(s_n)), \ \forall t \geq 0.$$

It follows that

$$0 = \lim_{n\to+\infty} d(y(t), x(t + s_n)) = \lim_{n\to+\infty} d(y(t), e^{t\odot A}(x(s_n))), \forall t \geq 0.$$

On the other hand, by *Theorem 4.3,(i)* we have

$$||e^{t\odot A}(x(s_n))||_{\mathcal{F}} \leq |||e^{t\odot A}|||_{\mathcal{F}}||x(s_n)||_{\mathcal{F}},$$

which implies

$$\lim_{n\to+\infty} ||e^{t\odot A}(x(s_n))||_{\mathcal{F}} = 0.$$

But $d(y(t), \tilde{0}_X) \leq d(y(t), e^{t\odot A}(s_n)) + d(e^{t\odot A}(s_n), \tilde{0}_X)$. Passing to limit with $n \to +\infty$, from the above relation we obtain

$$d(y(t), \tilde{0}_X) = 0, \forall t \in \mathbb{R}_+,$$

which immediately implies

$$y(t) = \tilde{0}_X, \forall t \in \mathbb{R}_+.$$

Now, for $t > M$ we get $t - s_n > 0$, which combined with

$$\lim_{n\to+\infty} d(y(t), x(t+s_n)) = \lim_{n\to+\infty} d(y(t-s_n), x(t)) = 0, t \in \mathbb{R}$$

and with $y(t) = \tilde{0}_X, \forall t \in \mathbb{R}_+$, immediately proves that

$$x(t) = \tilde{0}_X, \forall t > M.$$

From the above point (vii), it follows that

$$x(t) = \tilde{0}_X, \forall t \in \mathbb{R}.$$

(x) The proof is identical to the proof of *Theorem 2.1.10* [80], by using the fact that (X, d) is a complete metric space as well as the properties of d as a metric.   □

Remark 4.7. The hypothesis in the above *Theorem 4.6* (ix) is stronger than that in the case of Banach-space valued functions, where it is $\inf\{||x(t)||_{\mathcal{F}}; t \in \mathbb{R}\} = 0$.

This happens because in the case of $(X, \oplus, \odot, d)$ the property $T(t+s) = T(t)[T(s)]$ does not hold for all $t, s \in \mathbb{R}$ (it holds only for all $t, s \geq 0$ or for all $t, s < 0$ and does not hold if $ts < 0$).

Regarding the Hukuhara derivative of almost automorphic functions, we present the following result.

Theorem 4.8. *Assume that $f : \mathbb{R} \to \mathbb{R}_{\mathcal{F}}$ is almost automorphic and the Hukuhara derivative $f' : \mathbb{R} \to \mathbb{R}_{\mathcal{F}}$ exists and is uniformly continuous on $\mathbb{R}$.*

Then f' is almost automorphic.

Proof. Observe that for any $a, b \in \mathbb{R}$, with $a < b$, the Leibniz-Newton formula holds, i.e. $f(b) = f(a) \oplus \int_a^b f'(t)dt$, that is there exists $f(b) \ominus f(a) = \int_a^b f'(t)dt$. This implies

$$n \odot \int_0^{1/n} f'(t+s)ds = n \odot [f(t+1/n) \ominus f(t)], \int_9^{1/n} f'(t)ds$$

$$= \frac{1}{n} \odot f'(t).$$

We get

$$D(n \odot [f(t+1/n) \ominus f(t)], f') = D(n \odot \int_0^{1/n} f'(t+s)ds,$$

$$n \odot \int_0^{1/n} f'(t)ds)$$

$$= nD(\int_0^{1/n} f'(t+s)ds,$$

$$\int_0^{1/n} f'(t)ds)$$

$$\leq n \int_0^{1/n} D(f'(t+s), f'(t))ds.$$

The last inequality follows because the continuity of $f'(s)$ on $[0, \frac{1}{n}]$ implies the continuity of $F(s) = D(f'(t+s), f'(t))$ as function of s.

From the uniform continuity of f', for any $\varepsilon > 0$, there exists $\delta > 0$, such that for all $|s| < \delta, t \in \mathbb{R}$, we have

$$D(f'(t+s), f'(t)) < \varepsilon,$$

that is there exists n_0 such that for all $n \geq n_0$ we have $\frac{1}{n} < \delta$ and therefore

$$D(n \odot [f(t+1/n) \ominus f(t)], f'(t)) < \varepsilon, \forall t \in \mathbb{R}.$$

As a conclusion, the sequence

$$F_n(t) := n \odot [f(t+1/n) \ominus f(t)], n = 1, 2...,$$

converges uniformly on $\mathbb{R}$ to $f'(t)$.

If we prove that each function $F_n(t)$ is almost automorphic, then according to *Theorem 4.6* (x), will follow that f' is almost automorphic. For that, we need to prove the following helpful result :

If f_1, f_2 are almost automorphic and for all $t \in \mathbb{R}$ there exists $f_3(t) = f_1(t) \ominus f_2(t)$, then f_3 is almost automorphic.

Indeed, from hypothesis we have $f_1(t) = f_2(t) \oplus f_3(t), \forall t \in \mathbb{R}$, which according the definition of $\oplus$ means

$$[f_1(t)]^r = [f_2(t)]^r + [f_3(t)]^r, r \in [0,1], t \in \mathbb{R}$$

i.e.

$$[f_1(t)_-(r), f_1(t)_+(r)] = [[f_2(t)_-(r), f_2(t)_+(r)]$$
$$+ [f_3(t)_-(r), f_3(t)_+(r)], \forall r \in [0,1].$$

This last formula gives for all $r \in [0,1], t \in \mathbb{R}$

$$f_3(t)_-(r) = f_1(t)_-(r) - f_2(t)_-(r), f_3(t)_+(r)$$
$$= f_1(t)_+(r) - f_2(t)_+(r).$$

But from the definition of D, it easily follows that $f_1(t)$ is almost automorphic, if and only if all real-valued functions,

$$f_1(t)_-(r), f_1(t)_+(r), r \in [0,1]$$

are almost automorphic (as functions of t).

Similar result holds for $f_2(t)$.

Then from the classical theory we immediately get that all $f_3(t)_-(r), f_3(t)_+(r), r \in [0, 1]$ are almost automorphic as functions of t, which finally gives that $f_3(t)$ is almost automorphic.

The theorem is proved.   □

Regarding the integral of almost automorphic functions, we present

Theorem 4.9. *Let $f : \mathbb{R} \to X$ be almost automorphic and consider the function $F : \mathbb{R} \to X$ defined by $F(t) = \int_0^t f(s)ds$, where $(X, \oplus, \odot, d)$ is any of the spaces considered in Section 4.2.*

Then F is almost automorphic if and only if its range $R_F = \{F(t); t \in \mathbb{R}\}$ is relatively compact in X.

Proof. We adapt the proof of *Theorem 2.4.4* in [80] to our case. According to *Theorem 4.6 (v)*, it suffices to prove that if R_F is relatively compact, then F is almost automorphic.

Since f is almost automorphic and R_F is relatively compact in X, given (s_n'') a sequence of real numbers, there exists a subsequence $s_n')$ and $\alpha_1 \in X$ such that

$$\lim_{n \to +\infty} d(f(t + s_n'), g(t)) = \lim_{n \to +\infty} d(f(t), g(t - s_n'))$$
$$= \lim_{n \to +\infty} d(F(s_n'), \alpha_1)) = 0.$$

Then, as in [80], we get

$$F(t + s_n') = F(s_n') \oplus \int_0^t f(r + s_n')dr.$$

We have

$$\lim_{n \to +\infty} d(F(t + s_n'), \alpha_1 \oplus \int_0^t g(r)dr) = 0.$$

Indeed, denoting $g_n(r) = f(r + s'_n)$ obviously

$$\lim_{n \to +\infty} d(g_n(r), g(r)) = 0,$$

pointwise with respect to r and because f is almost automorphic, by *Theorem 4.6 (iv)*, f is bounded, i.e.

$$\sup\{||f(t)||_{\mathcal{F}}; t \in \mathbb{R}\} < M.$$

Denoting the real functions of real variable $h_n(r) = d(g_n(r), g(r))$, we have $\lim_{n \to +\infty} h_n(t) = 0$, for all $r \in \mathbb{R}$ and

$$h_n(r) \leq d(g_n(r), \tilde{0}_X) \oplus d(\tilde{0}_X, g(r))$$
$$= ||g_n(r)||_{\mathcal{F}} \oplus ||g(r)||_{\mathcal{F}} \leq 2||f||_{\mathcal{F}}$$
$$= 2M.$$

But

$$d(F(t + s'_n), \alpha_1 \oplus \int_0^t g(r)dr) = d(F(s'_n)$$
$$\oplus \int_0^t g_n(r)dr, \alpha_1 \oplus \int_0^t g(r)dr)$$
$$\leq d(F(s'_n), \alpha_1)$$
$$+ d(\int_0^t g_n(r)dr, \int_0^t g(r)dr)$$
$$\leq d(F(s'_n), \alpha_1)$$
$$+ \int_0^t d(g_n(r), g(r))dr,$$

which, by the Lebesgue's Dominated Convergence theorem, implies the convergence to zero of the last expression and therefore we get the required relation

$$\lim_{n \to +\infty} d(F(t + s'_n), \alpha_1 \oplus \int_0^t g(r)dr) = 0.$$

Note that the inequality

$$d\left(\int_0^t g_n(r)dr, \int_0^t g(r)dr\right) \le \int_0^t d(g_n(r), g(r))dr$$

used above is well-known in the case of fuzzy-number-valued functions. But it can similarly be extended to functions with values in $(X, \oplus, \odot, d)$, taking also into account that all the functions are continuous (actually it follows in an easy way from the definition of Riemann integral as limit of Riemann integral sums and the properties of metric d in X).

Now, denoting $G(t) := \alpha_1 \oplus \int_0^t g(r)dr$, from the relation

$$\lim_{n \to +\infty} d(F(t + s'_n), G(t)) = 0, \quad for\ all\ t \in \mathbb{R},$$

we get that the range of G also is relatively compact and the the following inequality

$$\sup\{\|G(t)\|_{\mathcal{F}}; t \in \mathbb{R}\} \le \sup\{\|F(t)\|_{\mathcal{F}}; t \in \mathbb{R}\}$$

holds.

So there is a subsequence $(s_n)_n$ of $(s'_n)_n$ and $\alpha_2 \in X$, such that

$$\lim_{n \to +\infty} d(G(-s_n), \alpha_2) = 0.$$

Then reasoning exactly as in [80],

we get

$$\lim_{n \to +\infty} d(G(t - s_n), \alpha_2 \oplus F(t)) = 0.$$

It remains to prove that $\alpha_2 = \tilde{0}_X$.

As in *Theorem 2.4.4* [80], we get

$A_s(F)(t) = \alpha_2 \oplus F(t)$, for all $t \in \mathbb{R}$, where we use the notations $s = (s_n)$, $A_s(F) \equiv T_s[T_{-s}(F)]$ and T_s is defined by $T_s(F) = H$, with H given by the relation $\lim_{n \to +\infty} F(t + s_n) = H(t), \forall t \in \mathbb{R}$.

Denoting $A_s^n := A[A_s^{n-1}]$, we get $A_s^n(F)(t) = n \odot \alpha_2 \oplus F(t)$. Firstly, let us prove

$$\sup\{\|A_s^n(F)(t)\|_{\mathcal{F}}; t \in \mathbb{R}\} \sup\{\|F(t)\|_{\mathcal{F}}; t \in \mathbb{R}\}.$$

For that, it suffices to prove the inequality

$$\sup\{\|A_s(F)(t)\|_{\mathcal{F}}; t \in \mathbb{R}\} \le \sup\{\|F(t)\|_{\mathcal{F}}; t \in \mathbb{R}\}.$$

In this sense we need the following result in $(X \oplus, \odot, d)$: if

$$\lim_{n \to +\infty} d(x_n, l) = 0,$$

then

$$\|l\|_{\mathcal{F}} = d(\tilde{0}_X, l) \le d(l, x_n) + d(x_n, \bar{0}_X),$$

where from passing to limit, we get

$$\|l\|_{\mathcal{F}} \le \lim_{n \to +\infty} \|x_n\|_{\mathcal{F}}.$$

Now, applying this result for $x_n = G(t - s_n)$ and $l = \alpha_2 \oplus F(t) = A_s(t)$, we obtain

$$\|A_s(F)(t)\|_{\mathcal{F}} \le \lim_{n \to +\infty} \|G(t - s_n)\|_{\mathcal{F}} \le \|F(t)\|_{\mathcal{F}}, \forall t \in \mathbb{R}.$$

Passing to supremum with $t \in \mathbb{R}$, we obtain the desired inequality. Note here that because the range of F is relatively compact, it follows that $\sup\{\|F(t)\|_{\mathcal{F}}; t \in \mathbb{R}\} < +\infty$.

Indeed, since R_F is bounded in the metric space, there exists $M > 0$ such that $d(x, y) \le M, \forall x, y \in R_F$. Then, we get

$$||F(t)||_{\mathcal{F}} = d(\tilde{0}_X, F(t)) \le d(\tilde{0}_X, F(t_0)) + d(F(t_0), F(t))$$

$$\le ||F(t_0)||_{\mathcal{F}} + M$$

$$< +\infty,$$

Finally, from $n \odot \alpha_2 = [A_s^n(F)(t) \oplus F(t)] \ominus F(t)$, we have

$$||n \odot \alpha_2||_{\mathcal{F}} = ||A_s^n(F)(t) \oplus F(t)||_{\mathcal{F}}$$

$$\le ||A_s^n(F)(t)||_{\mathcal{F}} + ||F(t)||_{\mathcal{F}}$$

$$\le 2||F(t)||_{\mathcal{F}},$$

where from passing to limit with $n \to +\infty$ we obtain a contradiction if $\alpha_2 \ne 0$.

Note that we used here the following inequality in $(X, \oplus, \odot, d)$:
if there exists $a \ominus b$, then

$$||a \ominus b||_{\mathcal{F}} = d(\tilde{0}_X, a \ominus b) = d(\tilde{0}_X \oplus b, (a \ominus b) \oplus b) =$$

$$d(b, a) \le d(b, \tilde{0}_X) + d(\tilde{0}_X, a) = ||a||_{\mathcal{F}} + ||b||_{\mathcal{F}}.$$

The theorem is thus completely proved. $\square$

Another useful result concerning the semigroups of operators is the following.

Theorem 4.10. . *Let* $x : \mathbb{R}_+ \to X$ *and* $f : \mathbb{R} \to X$ *be two continuous functions and* $T = (T(t))_{t \in \mathbb{R}_+}$ *be a* C_0-*semigroup of linear operators on* $(X, \oplus, \ominus, d)$. *Suppose that*

$$x(t) = T(t)(x(0)) \oplus \int_0^t T(t-s)(f(s))ds, t \in \mathbb{R}_+.$$

Then for t given in $\mathbb{R}$ *and* $b > a > 0$, $a + t > 0$, *we have*

$$x(t+b) = T(t+a)(x(b-a)) \oplus \int_{-a}^t T(t-s)(f(s+b))ds.$$

Proof. As in the proof of *Theorem 2.4.7* [80], we get

$$x(t+b) = T(t+a) \left[x(b-a) \ominus \int_0^{b-a} T(b-a-s)(f(s))ds \right]$$
$$\oplus \int_0^{t+b} T(t+b-s)(f(s))ds.$$

But if O is a linear operator on X and if there exists the quantity $x \ominus y$, then $O(x \ominus y) = O(x) \ominus O(y)$.

Indeed, if we denote $x \ominus y = a$, it follows $x = y \oplus a$ and $O(x) = O(y) \oplus O(a)$, which means $O(a) = O(x) \ominus O(y)$.

Then from the above relation we get

$$x(t+b) \ominus T(t+a) \left[\int_0^{b-a} T(b-a-s)(f(s))ds \right] = T(t+a)[x(b-a)] \oplus$$
$$\int_0^{t+b} T(t+b-s)(f(s))ds.$$

Taking into account that T commutes with the integral (since it is linear and continuous operator), by the property $T(u+v) = T(u)[T(v)], \forall u, v \in \mathbb{R}_+$ and making the substitution $u = s - b$, we obtain

$$x(t+b) \oplus \int_{-b}^{-a} T(t-u)[f(u+b)]du = T(t+a)[x(b-a)]$$
$$\oplus \int_{-b}^{t} T(t-u)[f(u+b)]du.$$

But because $t > -a$, we can write

$$\int_{-b}^{t} T(t-u)[f(u+b)]du = \int_{-b}^{-a} T(t-u)[f(u+b)]du$$
$$\oplus \int_{-a}^{t} T(t-u)[f(u+b)]du.$$

We then immediately get the required relation in the statement of theorem. Note that we used the following relation :

$$If \quad A \oplus E = B \oplus E, \quad then \quad A = B.$$

This is a trivial consequence of the relations $d(A, B) = d(A \oplus E, B \oplus E) = 0$, which implies $A = B$, since d is a metric.

The proof is achieved.   □

In the study of almost automorphic solutions of nonlinear fuzzy differential equations , the following concepts and results can be useful. We follow here the ideas in *Section 1.6.2*. Actually the results in the case of Banach spaces remain the same for the case of our metric spaces $(X, \oplus, \odot, d)$.

Definition 4.11. *A continuous function* $f : \mathbb{R} \times X \to X$ *is said to be almost automorphic in* $t \in \mathbb{R}$ *for each* $x \in X$, *if for every sequence of real numbers* (r_n), *there exists a subsequence* (s_n) *such that for all* $t \in \mathbb{R}$ *and* $x \in X$, *there exists* $g(t, x)$ *with the property*

$$\lim_{n \to +\infty} d(f(t + s_n, x), g(t, x)) = \lim_{n \to +\infty} d(g(t - s_n, x), f(t, x)) = 0.$$

The following simple properties hold.

Theorem 4.12. *(i) If* $f_1, f_2 : \mathbb{R} \times X \to X$ *are almost automorphic in* t *for each* $x \in X$, *then* $f_1 \oplus f_2$ *and* $c \odot f_1$, *where* $c \in \mathbb{R}$ *are also almost automorphic in* t *for each* $x \in X$.

(ii) If $f(t, x)$ *is almost automorphic in* t *for each* $x \in X$ *then* $\sup\{\|f(t, x)\|_{\mathcal{F}}; t \in \mathbb{R}\} < +\infty$. *Also, for the corresponding function* g *in Definition 4.11, we have* $\sup\{\|g(t, x)\|_{\mathcal{F}}; t \in \mathbb{R}\} < +\infty$.

(iii) If $f(t, x)$ is almost automorphic in t for each $x \in X$ and if

$$d(f(t, x), f(t, y)) \leq L \ d(x, y), \forall x, y \in X,$$

and $t \in \mathbb{R}$, where L is independent of x, y and t, then for the corresponding g in Definition 4.11, we have

$$d(g(t, x), g(t, y)) \leq L \ d(x, y), \forall x, y \in X$$

and $t \in \mathbb{R}$.

(iv) Let $f(t, x)$ be almost automorphic in t for each $x \in X$ such that $d(f(t, x), f(t, y)) \leq L \ d(x, y), \forall x, y \in X$ and $t \in \mathbb{R}$, where L is independent of x, y and t. If $\varphi : \mathbb{R} \to X$ is almost automorphic then the function $F : \mathbb{R} \to X$ defined by $F(t) = f(t, \varphi(t))$ is almost automorphic.

Proof. The proofs are similar to the proofs of the analogous Results in *Section 2.2* in [80] respectively, taking into account that everywhere in the proofs, the expressions of the type $\|x - y\|$ in the case of Banach spaces have to be replaced by $d(x, y)$ in the case of $(X, \oplus, \odot, d)$ spaces.

Similar to the classical case in *Sections 1.6.1 and 1.6.2*, the concept in *Definition 4.4* can be generalized as follows.

Definition 4.13. . *A continuous function $f : \mathbb{R}_+ \to X$ is said to be asymptotically almost automorphic if it admits the decomposition*

$$f(t) = g(t) \oplus h(t), \quad t \in \mathbb{R}_+,$$

where $g : \mathbb{R} \to X$ is almost automorphic and $h : \mathbb{R}_+ \to X$ is a continuous function with $\lim_{t \to +\infty} \|h(t)\|_{\mathcal{F}} = 0$.

Thus g and h are called the principal and the corrective terms

of f, respectively.

Remark 4.14. Every almost automorphic function restricted to $\mathbb{R}_+$ is asymptotically almost automorphic, by taking $h(t) = \tilde{0}_X, \forall t \in \mathbb{R}_+$.

Regarding this new concept, the following results in the classical case hold.

Theorem 4.15. . *Let f, f_1, f_2 be asymptotically almost automorphic. Then we have :*

(i) $f_1 \oplus f_2$ and $c \odot f, c \in \mathbb{R}$ are asymptotically almost automorphic ;

(ii) For fixed $a \in \mathbb{R}_+$, the function $f_a(t) = f(t + a)$ is asymptotically almost automorphic ;

(iii) f is bounded, i.e. $\sup\{\|f(t)\|_{\mathcal{F}}; t \in \mathbb{R}_+\} < +\infty$.

(iv) Let $(X, \oplus, \odot, d), (Y, \oplus, \odot, \rho)$ be any from the spaces considered in Section 4.2 and $f : \mathbb{R}_+ \to X$ be an almost automorphic function, $f = g \oplus h$. Let $\phi : X \to Y$ be continuous and assume there is a compact set B in (X, d) which contains the closures of $\{f(t); t \in \mathbb{R}_+\}$ and $\{g(t); t \in \mathbb{R}_+\}$. If, in addition, for all $t \in \mathbb{R}_+$ there exists continuous $\Gamma(t) = \phi(f(t)) \ominus \phi(g(t))$, then $\phi \circ f : \mathbb{R}_+ \to Y$ is asymptotically almost periodic ;

(v) In general, the decomposition of an asymptotically almost automorphic function is not unique.

Proof. (i) and (ii) are immediate from *Definition 4.13* and *Theorem 4.6, (i),(ii),(iii).*

(iv) We have $\phi(f(t)) = \phi(g(t)) \oplus \Gamma(t)$, where by *Theorem 4.6, (viii)* , $\phi(g(t))$ is almost automorphic. Therefore, it remains to prove that $\lim_{t \to +\infty} ||\Gamma(t)||_{\mathcal{F}} = 0$.

We get

$$||\Gamma||_{\mathcal{F}} = \rho(\tilde{0}_Y, \phi(f(t)) \ominus \phi(g(t)))$$
$$= \rho(\phi(g(t)), \phi(f(t)) \ominus \phi(g(t)) \oplus \phi(g(t)))$$
$$= \rho(\phi(g(t)), \phi(f(t))).$$

For the rest of the proof, we follow the lines as in the proof of *Theorem 2.5.7* in [80].

Here we have used the property that

If $a, b \in X$ are such that $a \ominus b$ exists, then $(a \ominus b) \oplus b = a$.

This is atraightforward. Indeed, denoting $x = a \ominus b$, by definition we get $a = x \oplus b$, which implies $(a \ominus b) \oplus b = [(x \oplus b) \ominus b] \oplus b = x \oplus b = a$, because it is evident that $(x \oplus b) \ominus b = x$.

(v) First, we prove the following relation : if $a, b, c \in X$ are such that there exists $b \ominus c$, then $a \oplus (b \ominus c) = (a \oplus b) \ominus c$. Indeed, denoting $x = b \ominus c$, by definition we get $b = x \oplus c$ and therefore $(a \oplus b) \ominus c) = (a \oplus x \oplus c) \ominus c = a \oplus x = a \oplus (b \ominus c)$, because, in general it is immediate that $(A \oplus B) \ominus B = A$. Note that this relation does not hold if $b \ominus c$ does not exist.

Let us suppose that $f : \mathbb{R}_+ \to X$ admits two decompositions

$$f(t) = g_i(t) \oplus h_i(t), t \in \mathbb{R}_+, i = 1, 2.$$

We get

$$g_1(t) \oplus h_1(t) = g_2(t) \oplus h_2(t),$$

which implies

$$g_1(t) = [g_2(t) \oplus h_2(t)] \ominus h_1(t)$$

and $h_2(t) = [h_1(t) \oplus g_1(t)] \ominus g_2(t)$, for all $t \in \mathbb{R}_+$.

Now, let us suppose that for each $t \in \mathbb{R}_+$ there exists $h_2(t) \ominus h_1(t)$ or $g_1(t) \ominus g_2(t)$.

Then by the above relation we get $g_1(t) = g_2(t) \oplus [h_2(t) \ominus h_1(t)]$ or $h_2(t) = h_1(t) \oplus [g_1(t) \ominus g_2(t)]$, respectively. Both cases imply the same relation $g_1(t) \ominus g_2(t) = h_2(t) \ominus h_1(t), \forall t \in \mathbb{R}_+$. In this case, we obtain

$$d(g_1(t) \ominus g_2(t), \tilde{0}_X) = d(h_2(t) \ominus h_1(t), \tilde{0}_X) \leq$$

$$d(h_2(t), \tilde{0}_X) + d(\tilde{0}_X, h_1(t)) = ||h_2(t)||_{\mathcal{F}} + ||h_1(t)||_{\mathcal{F}}.$$

Passing to limit with $t \to +\infty$, we get

$$\lim_{t \to +\infty} d(g_1(t) \ominus g_2(t), \tilde{0}_X) = 0.$$

According to the proof of *Theorem 4.6*, $g_1(t) \ominus g_2(t)$ is almost automorphic. Considering the sequence $n, n = 1, 2, ...,$, there exists a subsequence (n_k) such that

$$\lim_{k \to +\infty} d(g_1(t + n_k) \ominus g_2(t + n_k), F(t)) =$$

$$\lim_{k \to +\infty} d(g_1(t) \ominus g_2(t), F(t - n_k)) = 0,$$

pointwise on $\mathbb{R}$.

Using the inequality

$$d(F(t), \tilde{0}_X) \leq d(g_1(t) \ominus g_2(t), \tilde{0}_X) + d(g_1(t + n_k) \ominus g_2(t + n_k), F(t)),$$

and passing to limit with $k \to +\infty$, it follows that $F(t) = \tilde{0}_X, \forall t \in$ $\mathbb{R}$ and consequently $g_1(t) \ominus g_2(t) = \tilde{0}_X, \forall t \in \mathbb{R}$. Therefore, $g_1 \equiv g_2$ and $h_1 \equiv h_2$.

As a conclusion, the uniqueness follows only in the special case when for each $t \in \mathbb{R}_+$ there exists $h_2(t) \ominus h_1(t)$ or $g_1(t) \ominus g_2(t)$. But in general, this condition does not hold, which implies that the uniqueness of decomposition does not hold. $\square$

Remark 4.16. 1) In comparison with the case of Banach-space valued functions, in *Theorem 4.15, (iv)* we need the additional hypothesis that for all $t \in \mathbb{R}_+$, there exists (continuous) $\Gamma(t) = \phi(f(t)) \ominus \phi(g(t))$.

2) The effect of *Theorem 4.15, (v)*, is that it produces less properties than in the classical case.

In what follows, we are concerned with the asymptotical behavior of asymptotically almost automorphic semigroups of linear operators $T = T(t), t \in \mathbb{R}_+$ on $(X, \oplus, \odot, d)$. We present some topological and asymptotic properties based on the Nemytskii and Stepanov theory of dynamical systems as presented in *Section 1.6.2.*

Definition 4.17. . *A mapping* $u : \mathbb{R}_+ \times X \to X$ *is called a dynamical system if :*

(i) $u(\tilde{0}_X, x) = x, \forall x \in X$ *;*

(ii) $u(., x) : \mathbb{R}_+ \to X$ *is continuous for any* $t > 0$ *and right-continuous at* $t = 0$, *for each* $x \in X$. *(The mapping* $u(., x)$ *is called a motion originating at* $x \in X$*).*

(iii) $u(t,.): X \to X$ is continuous for each $t \geq 0$;

(iv) $u(t+s,x) = u(t,u(s,x)), \forall x \in X, t,s \in \mathbb{R}_+$.

We now present the following important correspondence.

Theorem 4.18. . *Every C_0-semigroup $(T(t))_{t\geq 0}$ on $(X, \oplus, \odot, d)$ determines a dynamical system and conversely, by defining $u(t,x) = T(t)(x), t \in \mathbb{R}_+, x \in X$.*

Proof. Similar to *Theorem 1.43*.

In the rest of this section, $T = (T(t))_{t\in \mathbb{R}_+}$ will be a C_0-semigroup of linear operators on $(X, \oplus, \odot, d)$ such that for fixed $x_0 \in X$, the motion $T(t)(x_0) : \mathbb{R}_+ \to X$ is an asymptotically almost automorphic function in the sense of *Definition 4.13*, with principal term f and corrective term h.

Definition 4.19. . *A function $\varphi : \mathbb{R} \to X$ is said to be a complete trajectory of T if it satisfies the functional equation*

$$\varphi(t) = T(t-a)(\varphi(a)),$$

for all $a \in \mathbb{R}, t \geq a$.

We have the following

Theorem 4.20. . *The principal term f of $T(t)(x_0)$ is a complete trajectory for T.*

Proof Similar to the proof of *Theorem 1.45*.

Definition 4.21.

$$\omega^+(x_0) = \{y \in X; \exists 0 \le t_n \to +\infty, \lim_{n \to +\infty} d(T(t)(x_0), y) = 0\}$$

is called the ω-limit set of $T(t)(x_0)$.

$$\omega_f^+(x_0) = \{y \in X; \exists 0 \le t_n \to +\infty, \lim_{n \to +\infty} d(f(t_n), y) = 0\}$$

is called the ω-limit set of f, the principal term of $T(t)(x_0)$.

$\gamma^+(x_0) = \{T(t)(x_0); t \in \mathbb{R}+\}$ is the trajectory of $T(t)(x_0)$.

A set $B \subseteq X$ is said to be invariant under the semigroup $T = (T(t))_{t \in \mathbb{R}_+}$, if

$$T(t)(y) \in B, \forall y \in B, t \in \mathbb{R}_+.$$

$e \in X$ is called a rest-point for the semigroup T if

$$T(t)(e) = e, \forall t \ge 0.$$

The following properties hold.

Theorem 4.22. . *(i) $\omega^+(x_0)$ is not empty, $\omega^+(x_0) = \omega_f^+(x_0)$, $\omega^+(x_0)$ is invariant under T and is closed in X, $\omega^+(x_0)$ is compact if $\gamma^+(x_0)$ is relatively compact. Also, if x_0 is a rest-point of the semigroup T then $\omega^+(x_0) = \{x_0\}$.*

(ii) If we denote $\gamma_f(x_0) = \{f(t); t \in \mathbb{R}\}$ then $\gamma_f(x_0)$ is relatively compact and invariant under the semigroup T.

(iii) If we denote $\nu(t) = \inf\{d(T(t(x_0), y); y \in \omega^+(x_0)\}$, then $\lim_{t \to +\infty} \nu(t) = 0$.

Remark 4.23. All the above results show us that the properties in the cases of Banach spaces, remain valid for the case of the more general $(X, \oplus, \odot, d)$ spaces.

4.4 Applications to Fuzzy Differential Equations

It is known that the classical (abstract) differential equations ,i.e. whose solutions are real-valued functions (or Banach-space valued functions, respectively) often represent an idealization of real situations, where imprecision may in fact play a significant role. A way to solve this shortcoming is to consider random differential equations (i.e. whose solutions are random-variable-valued functions), which have been used to incorporate the effects of statistical fluctuations.

On the other hand, imprecision due to uncertainty or vagueness suggests the introduction of so-called fuzzy differential equations, i.e whose solutions represent functions with values in $\mathbb{R}_{\mathcal{F}}$ or more general, with values in X, where $(X, \oplus, \odot, d)$ represents any from the spaces introduced in *Section 4.2*. Applications of the semigroup of operators in solving fuzzy partial differential equations have been done in the recent paper [38].

Now we like to illustrate the idea of propagation of almost automorphy from the fuzzy input data to the solutions of fuzzy differential equations.

The first result in this sense is the following.

Theorem 4.24. . *Let us consider the fuzzy differential equation*

$$y'(t) \oplus y(t) = g(t),$$

where $y'(t)$ means the Hukuhara derivative , $g : \mathbb{R} \to \mathbb{R}_{\mathcal{F}}$ is of the particular form $g(t) = c \odot f(t), \forall t \in \mathbb{R}$, with $c \in \mathbb{R}_{\mathcal{F}}$ a fuzzy number and $f : \mathbb{R} \to \mathbb{R}$ a usual almost automorphic function.

In addition suppose that f satisfies the condition $f(t) \geq 0, \forall t \in \mathbb{R}$.

Then the function defined by $y : \mathbb{R} \to \mathbb{R}_{\mathcal{F}}$ defined by

$$y(t) = c \odot \int_{-\infty}^{t} e^{u-t} f(u) du,$$

is a almost automorphic function on $\mathbb{R}$ and satisfies the above fuzzy differential equation for all $t \in \Omega$, where

$$\Omega = \{t \in \mathbb{R}; f(t) > \int_{-\infty}^{t} e^{u-t} f(u) du\}.$$

Proof. First we notice that by the hypothesis on f and by *Definition 4.4*, it immediately follows that $g : \mathbb{R} \to \mathbb{R}_{\mathcal{F}}$ is almost automorphic.

Now, according to [38], *Section 4, Theorem 14*, $y(t)$ satisfies the fuzzy equations for all $t \in \Omega$.

Then the function $F(t) = \int_{-\infty}^{t} e^{u-t} f(u) du$ is almost automorphic on $\mathbb{R}$, which by the *Definition 4.4* and the properties of D immediately implies that $y(t) = c \odot F(t)$ is almost automorphic.

Remark 4.25. Theorem 4.24 remains valid for the more general differential equation

$$y'(t) \oplus y(t) = g(t),$$

where $y'(t)$ means the Hukuhara derivative , $g : \mathbb{R} \to X$ is of the particular form $g(t) = c \odot f(t), \forall t \in \mathbb{R}$, with $c \in X$ and $f : \mathbb{R} \to \mathbb{R}$ a usual almost automorphic function, and $(X, \oplus, \odot, d)$ is any from the spaces considered by the previous sections.

Before to consider the next result, let us make some remarks on the concepts of differentiability presented in *Section 4.2.* Firstly, note that the Hukuhara differentiability has the following shortcoming : if c is a fuzzy number, f is a real valued function (of real variable) differentiable on t and $g(t) = c \odot f(t)$ then for $f'(t) > 0$ we have $g'(t) = c \odot f'(t)$, while for $f'(t) < 0$, the function g is not Hukuhara differentiable on t.

This shortcoming is solved by the generalized differentiability, which at its turn has another shortcoming : if $f'(t) = 0$, then g is not necessarily generalized differentiable on t.

In the paper [9] these kinds of shortcomings are completely solved in such a way that g is always differentiable (in the new sense), if f is differentiable. The new concept of differentiability can be stated as follows (see [9]) :

Let $f : (a,b) \to \mathbb{R}_{\mathcal{F}}$ be and $t \in (a,b)$. For a sequence of real numbers $h_n \searrow 0$, let us consider the sets

$$A_p^{(1)} = \{n \geq p; \exists E_n^{(1)} = f(t + h_n) \ominus f(t)\},$$

$$A_p^{(2)} = \{n \geq p; \exists E_n^{(2)} = f(t) \ominus f(t + h_n)\},$$

$$A_p^{(3)} = \{n \geq p; \exists E_n^{(3)} = f(t) \ominus f(t - h_n)\},$$

$$A_p^{(4)} = \{n \geq p; \exists E_n^{(4)} = f(t - h_n) \ominus f(t)\}.$$

We say that f is weakly generalized differentiable on t, if for any sequence $h_n \searrow 0$, there exists $p \in \mathbb{N}$ such that

$$A_p^{(1)} \bigcup A_p^{(2)} \bigcup A_p^{(3)} \bigcup A_p^{(4)} = \{n \in \mathbb{N}; n \geq p\}$$

and moreover, there exists an element in $\mathbb{R}_{\mathcal{F}}$ denoted by $f'(t)$, such that if for some $j \in \{1, 2, 3, 4\}$ we have $card(A_p^j) = +\infty$, then

$$\lim_{n \to +\infty, h_n \in A_p^j} D\left((-1)^{j+1} \odot E_n^j, f'(t)\right) = 0.$$

Remark 4.26. Obviously the generalized differentiability implies the weakly generalized differentiability but the converse is not true.

Now we are in position to present the following.

Theorem 4.27. . *Let us consider the fuzzy wave equation*

$$\frac{\partial^2 u(x,t)}{\partial x^2} = \frac{1}{c^2} \frac{\partial^2 u(x,t)}{\partial t^2}, t \geq 0, x \in \mathbb{R},$$

with the boundary conditions

$$u(x,0) = \alpha \odot f(x), \frac{\partial u(x,0)}{\partial t} = kg(x) \odot \alpha, x \in \mathbb{R}$$

where $\alpha \in \mathbb{R}_{\mathcal{F}}$, $c, k \in \mathbb{R}, c > 0$ and $f, g : \mathbb{R} \to \mathbb{R}$, with f of C^2-class and g of C^1-class. Here the differentiability is considered in the weak generalized sense.

Then

$$u(x,t) = \alpha \odot \left\{ [f(x-ct) + f(x+ct)]/2 + \frac{k}{2c} \int_{x-ct}^{x+ct} g(s)ds \right\}$$

satisfies the above fuzzy differential equation, and if in addition, f and g are almost automorphic, and $F(x) = \int_0^x g(s)ds$ is bounded on $\mathbb{R}$, then for each fixed $t \geq 0$, the solution $u(.,t)$ is almost automorphic.

Proof. It is evident from the classical theory that the function between the brackets in the above expression of $u(x,t)$ is almost automorphic with respect to the variable x (for each fixed t), which by *Definition 4.4* immediately implies that $u(x,t)$ also is almost automorphic with respect to x.

Also, the fact that $u(x, t)$ satisfies the above fuzzy wave equation is immediate from the classical theory and from the property of the above weak generalized differentiability.

Remark 4.28. According to the above *Theorem 4.3*, the so-called mild solution

$$u(t) = T(t)(u_0) \oplus \int_0^t T(t - s)g(s)ds,$$

where $T(t)$ represents the exponential operators in the same *Theorem 4.3, (iii)*, satisfies the abstract "fuzzy" differential equation

$$u'(t) = A[u(t)] \oplus g(t),$$

where u' is considered in the generalized sense.

4.5 Bibliographical Remarks and Open Problems

This chapter is based on a study by S. G. Gal and G. M. N'Guérékata ([41]).

It would be of interest to see for a Bohr-Neugebauer-N'Guérékata-type result (see e.g. [30]) in the case discussed in the last section above, i.e. how almost automorphy of the forced term along with other conditions, on the operator for instance, would produce almost automorphy of the solution to the fuzzy differential equation.

Also extensions to fuzzy settings of the results in *Theorems 2.17* and *Theorem 2.18* were obtained by C. S. Gal, S. G. Gal and G. N. N'Guérékata ([42]).

References

1. A. Alexiewicz, *Functional Analysis*, PWN, Warsaw, 1969.

2. L. Amerio and G. Prouse, *Almost Periodic Functions and Functional Equations*, Van Nostrand Reinhold, Co. New York-Toronto, 1971.

3. G. A. Anastassiou and S. Gal, *On a Fuzzy Trigonometric Approximation Theorem of Weierstrass-Type*, J. Fuzzy Math., 9(2001), No. 3, 701-708.

4. R. B. Basit, *Generalization of two theorems of M. I. Kadets concerning inetgral of abstract almost periodic functions*, Matematicheskie Zameki, Vol. **9**, No. 3,(March 1971) pp. 311-321.

5. R. B. Basit and L. Tsend, *The generalized Bohr-Neugebauer Theorem*, Differentsial'nye Uravneiya, Vol. **8**, No. 8, (August 1972), pp. 1343-1348.

6. R. B. Basit, *The relationship between almost-periodic Levitan functions and almost-automorphic functions*, Vestnik Moskovskogo Universiteta. Matematika, Vol. **26**, No. 4, (1971), pp. 11-15.

7. B. Basit, *Les Fonctions Abstraites Presque Automorphiques et Presque Périodiques au sens de Levitan, et leurs Differences*, Bull. Sc. Math. 2e Serie, **101**, (1977), 131-148.

8. B. Basit and M. Emam, *Differences of Functions in Locally Convex Sapces and Applications to Almost Periodic and Almost Automorphic Functions*, Annales Polonici Math. XLI (1983), 193-201.

9. B. Bede and S. G. Gal, *Generalizations of the Differentiability of Fuzzy-Number-Valued Functions with Applications to Fuzzy Differential Equations*, Fuzzy Stes and Systems, accepted.

10. S. Bochner, *Continuous Mappings of Almost Automorphic and Almost Periodic Functions*, Proc. Nat. Acad. Sci. USA **52** (1964), pp. 907–910.

11. S. Bochner, *Uniform Convergence of Monotone Sequences of Functions*, Proc. Nat. Acad. Sci. USA **47** (1961), pp. 582–585.

12. S. Bochner, *A new Approach to Almost-Periodicity*, Proc. Nat. Acad. Sci. USA **48** (1962), pp. 2039–2043.

13. S. Bochner and J. Von Neumann, *On Compact Solutions of Operational-Differential Equations*, I, Ann. Math **36** (1935), pp. 255–290.

14. H. Bohr, *Almost Periodic Functions*, Chelsea Publishing Company, New York, 1947.

15. D. Bugajewska, D. Bugajewski, *On Topological Properties of Solutions Sets for Differential Equations in Locally Convex Spaces*, Nonl. Anal. T. M. A., **47** (2001), 1211-1220.

16. D. Bugajewski and G. M. N'Guérékata, *On the Topological Structure of Almost Automorphic and Asymptotically Almost Automorphic Solutions of Differential and Integral Equations in Abstract Spaces*, preprint.

17. D. Bugajewski and G. M. N'Guérékata, *Almost Periodicity in Fréchet Spaces*, J. Math. Analysis and Appl., accepted.

18. D. Bugajewski, *On the Existence of Weak Solutions of Integral Equations in Banach Spaces*, Comment. Math. Univ. Carolinae **35** (1994), 35-41.

19. T. Cazenave and A. Haraux, *An Introduction to Semilinear Evolution Equations*, Oxford Lecture Series in Math. and its Appl. **13** Clarendon Press Oxford, 1998.

20. W. Congxin and G. Zengtai, *On Henstock Integral of Fuzzy-Number-Valued Functions*, *I*, Fuzzy Sets and Systems, 120(2001), 523-532.

21. W. Congxin, W. Guixiang, *The Integral over a Directed Line Segment of Fuzzy Mapping from the Fuzzy Number Space E into E and its Applications*, submitted.

22. W. A. Coppel, *Dichotomies in Stability Theory*, Springer-Verlag, Berlin-Heidelberg-New York, 1978.

23. C. Corduneanu, *Almost Periodic Functions*, Chelsea Publishing Company, New York, 1989.

24. C. Corduneanu and J. A. Goldstein, *Almost Periodicity of Bounded Solutions to Nonlinear Abstract Equations*, Diff. Eq., North-Holland Mathematics Studies **92** (1984), pp.115–121.

25. K. deLeeuw and I. Glicksberg, *Applications of Almost Periodic Compactifications*, Acta Math. **105** (1961), pp. 63–97.

26. T. Diagana, *Schrödinger Operators with Singular Potential*, Intern. J. Math. and Math. Sci. Vol. **29**, No. 6, (2002), 371-373.

27. T. Diagana, *A Generalization related to Schrödinger Operators with a Singular Potential*, Intern. J. Math. and Math. Sci., Vol. **29**, No. 10, (2002), 600-611.

28. T. Diagana, *Some Remarks on some Second-Order Hyperbolic Differential Equations*, Semigroup Forum, **68**, (2004), pp. 357-364.

29. T. Diagana and G. M. N'Guérékata, *Some Remarks on Almost Automorphic Soutions of some Abstract Differential Equations*, Far East J. Math. **8** (3) (2003), pp. 313-322.

30. T. Diagana and G. M. N'Guérékata, *On the Bohr-Neugebauer-NGuérékata Theorem*, J. of. Anal. Appl., Vol. **2** (2004), No.1, pp 1-10

31. T. Diagana and G. M. N'Guérékata, *On some Perturbations of some Abstract Differential Equations*, Comment. Math. Vol. **XLIII**, No. 2 (2003), pp. 201-206.

32. T. Diagana, G. M. N'Guérékata and N. V. Mink, *Almost Automorphic Solutions of Evolution Equations*, Proceedings of the Amer. Math. Soc., accepted.

33. C. Dunford and J. T. Schwartz, *Linear Operators*, Vol. **I**, Interscience, New York, 1969.

34. R. Dragoni, J. W. Macki, P. Nistri and P. Zecca, *Solution Sets of Differential Equations in Abstract Spaces*, Pitman Research Notes in Mathematics, vol. 342 (1996).

35. J. E. Egawa, *Eigenvalues of Some Almost Automorphic Functions*, Proc. Japan Acad. Ser. A, Math. Sci. **61** (1985), no. 7, 203-206.

36. A.M. Fink, *Almost Periodic Differential Equations*, Lecture Notes in Math., vol 37, Springer-Verlag, Berlin-Heidelberg-New York, 1974.

37. M. Fréchet, *Fonctions Asymptotiquement Presque Périodiques*, Revue Scientifique (Revue Rose Illustrée) **79** (1941), pp. 341-354.

38. C. S. Gal and S. G. Gal, *Semigroups of Operators on Spaces of Fuzzy-Number-Valued Functions with Applications to Fuzzy Differential Equations*, submitted.

39. S. G. Gal, *Approximation Theory in Fuzzy Setting*, Chapter 13 in Handbook of Analytic-Computational Methods in Applied Mathematics (ed. G.A. Anastassiou) Chapman & Hall/CRC, Boca Raton-London-New York-Washington D.C., 2000, pp.617-666.

40. S. G. Gal and B. Bede *Almost Periodic Fuzzy-Number-Valued Functions*, Fuzzy Sets and Systems, accepted.

41. S. G. Gal and G. M. N'Guérékata, *Almost Automorphic Fuzzy-Number-Valued Functions*, submitted.

42. C. S. Gal, S. G. Gal, and G. M. N'Guérékata, *Existence and Uniqueness of Almost Automorphic Mild Solutions to Some Semilinear Fuzzy Differential Equations* . Advances in Differential Equations. African Diaspora Journal of Mathematics, accepted.

43. S. Goldberg, *Unbounded Linear Operators, Theory and Applications*, McGraw-Hill Book Co., New York-San-Francisco-London, 1966.

44. J. A. Goldstein, *Convexity, Boundedness and Almost Periodicity for Differential Equations in Hilbert Spaces*, Intern. J. Math and Math. Sci. **2** (1979), pp. 1–13.

45. J. A. Goldstein, *Semigroups of Linear Operators and Applications*, Oxford University Press, Oxford, 1985.

46. J. A. Goldstein and G. M. N'Guérékata, *Almost Automorphic Solution of Semilinear Evolution Equations*, Proc. Amer. Math. Soc., to appear.

47. J. K. Hale, *Ordinary Differential Equations*, Wiley-Interscience, New York, 1969.

48. A. E. Hamza and G. L. Muraz, *Spectral Criteria of Abstract Functions; Integral and Difference Problems*, Acta Math. Vietnamica, **23** (1) (1998), pp. 171-184.

49. E. Hille and R. S. Phillips, *Functional Analysis and Semigroups*, Amer. Math. Soc. Coll. Publ., vol. XXXI (1957), Providence, RI.

50. Y. Hino, T. Naito, N. V. Minh and J. S. Shin, *Almost Periodic Solutions of Differential Equations in Banach Spaces*, Taylor and Francis, London-New-York, 2002.

51. Y. Hino and S. Murakami, *Almost Automorphic Solutions for Abstract Functional Differential Equations*, J. Math. Analysis and Appl.

52. V. M. Hokkanen and G. Morosanu, *Functional Methods in Differential Equations*, Chapman-Hall-CRC **432** Boca Raton-London-New York, 2002.

53. M. I. Kadets, *The Integration of Almost Periodic Functions with Values in a Banach Space*, Functional Analysis and its Applications **3** (1969), pp. 228–230.

54. J. Kopel, *On Vector-Valued Almost Periodic Functions*, Ann. Sc. Polon. Math. **25** (1952), pp. 100–105.

55. R. Larsen, *Functional Analysis*, Decker Inc. New York, 1973.

56. B. M. Levitan and V. V. Zhikov, *Almost Periodic Functions and Differential Equations* Cambridge University Press, Cambridge-London-New York, 1982.

57. F. X. Lin. *The Existence of Almost Automorphic Solution of Almost Automorphic Systems*, Ann. Diff. Eq. **3** (1987), no. 3, 329-349.

58. J. Liu, G. M. N'Guérékata and NGuyen V. Minh, *Almost Automorphic Solutions of Second Order Evolution Equations*, Appl. Analysis, to appear.

59. J. Locker, *Spectral Theory of Non-Self-Adjoint Two-Point Differential Operators*, AMS Mathematical Surveys and Monographs, Vol. **73**, (2000).

60. R. E. Megginson, *An Introduction to Banach Space Theory*, Graduate texts in Mathematics, **183**, Springer-Verlag, New-York, 1998.

61. V. Nemytskii and V. V. Stepanov, *Quality Theory of Differential Equations*, Princeton University Press, 1960.

62. G. M. N'Guérékata, *Almost Automorphic Functions and Applications to Abstract Evolution Equations*, Contemporary Math., Amer. Math. Soc. **252** (1999), pp. 71-76.

63. G. M. N'Guérékata, *Almost Automorphic Solutions of some Differential Equations in Banach Spaces*, Int. J. Math.-. Math. Sci. **23** (2000), pp. 361–365.

64. G. M. N'Guérékata, *An asymptotic Theorem for Abstract Differential Equations*, Bull. Australian Math. Soc. **33** (1986), pp. 139–144.

65. G. M. N'Guérékata, *On Almost Automorphic Differential Equations in Banach Spaces*, Pan American Math. J. **9** (1999), pp. 103–108.

66. G. M. N'Guérékata, *Quelques Remarques sur les Fonctions Asymptotiquement Presque-Automorphes*, Ann. Sci. Math Quebec **VII** (1983), pp. 185–191.

67. G. M. N'Guérékata, *Some Remarks on Asymptotically Almost Automorphic Functions*, Riv. di Mat. della Universita di Parma (4) **13** (1987), pp. 301–303.

68. G. M. N'Guérékata, *Sur les Solutions Presque Automorphes d'Equations Différentielles Abstraites*, Ann. Sci. Math Quebec **5** (1981), pp. 69–79.

69. G. M. N'Guérékata, *Almost Periodicity in Linear Topological Spaces and Applications to Abstract Differential Equations*, Int'l. J of Math. and Math Sci. **7** (1984), 529–540.

70. G. M. N'Guérékata, *Notes on Almost Periodicity in Topological Vector Spaces*, Int'l. J. of Math. and Math. Sci. **9** (1986), 201–206.

71. G. M. N'Guérékata, *Almost Periodic Solutions of certain Differential Equations in Fréchet Spaces*, Riv. di Mat. della Universita di Parma (5) **2** (1993), pp. 301–303.

72. G. M. N'Guérékata, *Remarques sur les Solutions Presque Périodiques de l'Equation $(d/dt - A)x = 0$*, Can. Math. Bull. **25** (1982), pp. 121–123.

73. G. M. N'Guérékata, *Almost Periodicity of some Solutions to Linear Abstract Equations*, Libertas Mathematica **XVI** (1996), 145–148.

74. G. M. N'Guérékata, *On Almost Periodic Solutions of the Differential Equation $x''(t) = Ax(t)$ in Hilbert Spaces*, Intern. J. Math. and Math. Sci. (2001), Vol. 28 (4), 247-249.

75. G. M. N'Guérékata, *Existence and Uniqueness of Almost Automorphic Mild Solutions to some Semilinear Abstract Differential Equations*, Semigroup Forum, to appear.

76. G. M. N'Guérékata, *Almost Automorphy and Almost Periodicity of Motions in Banach Spaces*, Forum Math.**13** (2000), pp. 581-588.

77. G. M. N'Guérékata, *On Weak-Almost Periodic Optimal Mild Solutions of some Linear Abstract Differential Equations*, Dynamical Systems and Differential Equations, A supplement volume to Discrete and Continuous Dynamical Systems, (2003), pp. 672-677.

78. G. M. N'Guérékata, *Notes on certain Almost Automorphic Abstract Differential Equations*, Far East J. Math. **12** (1) (2004), pp. 17-21.

79. G. M. N'Guérékata, *Remarks on Almost Automorphic Differential Equations*, Dynamical systems and differential equations (Kennesaw, GA, 2000). Discrete Cont. Dynam. Systems 2001, Added Volume, pp. 247-249.

80. G. M. N'Guérékata, *Almost Automorphic and Almost Periodic Functions in Abstract Spaces*, Kluwer Academic Publishers, 2001, New York-London-Moscow.

81. A. Pazy, *Semigroups of Linear Operators and Applications to Partial Differential Equations*, Appl. Math. Sci. **44**, Springer-Verlag, New York, 1983.

82. A. S. Rao, *On almost Automorphic Solution of Certain Abstract Differential Equations*, Indian J. Math. **33** (1991), 179-187.

83. M. Renardy and R. C. Rogers, *An Introduction to Partial Differential Equations*, Texts in Appl. Math. **13** (1992).

84. A. P. Robertson and W. Robertson, *Topological Vector Spaces*, Cambridge University Press, 1973.

85. W. M. Ruess and W.H. Summers, *Asymptotic Almost Periodicity and Motions of Semigroups of Operators*, Linear Algebra and its Applications **84** (1986), pp. 335–351.

86. M. Schechter, *Principles of Functional Analysis*, Academic Press, New York, 1973.

87. G. R. Sell and Y. You, *Dynamics of Evolutionary Equations*, Springer,**143** New York, 2002.

88. W. Shen and Y. Yi, *Almost Automorphic and Almost Periodic Dynamics in Skew-Product Semiflows*, Memoirs of the Amer. Math. Soc., **647** vol. 136, 1998.

89. W. A. Veech, *Almost Automorphic Functions on Groups*, Amer. J. Math. **87** (July 1965), pp. 719–751.

90. K. Yosida, *Functional Analysis*, Springer-Verlag, 1968.

91. T. Yoshizawa, *Stability Theory and the Existence of Periodic Solutions and Almost Periodic Solutions*, Springer-Verlag, New York-Heidelberg-Berlin, 1975.

92. S. Zaidman, *Almost Automorphic Solutions of some Abstract Evolution Equations*, Istituto Lombardo di Sci. e Lett. **110** (1976), pp. 578–588.

93. S. Zaidman, *Almost Periodic Functions in Abstract Spaces*, Pitman Advanced Publishing Program, Boston-London-Melbourne, 1985.

94. S. Zaidman, *Abstract Differential Equations*, Pitman Publishing Limited, San-Franciso-London-Melbourne, 1979.

95. S. Zaidman, *Behavior of Trajectories of C_0-semigroups*, Istituto Lambardo, Acc. Sci. Lett. Rend. A **114** (1980-82), pp. 205–208.

96. S. Zaidman, *Behavior of Trajectories of C_0-semigroups (II)*, Ann. Sc. Math., Quebec **6** (1982), pp. 215–220.

97. S. Zaidman, *Existence of Asymptotically Almost Periodic and of Almost Automorphic Solutions for some Classes of Abstract-Differential Equations*, Ann. Sc. Math., Quebec, **13** (1989), pp. 79–88.

98. S. Zaidman, *Topics in Abstract Differential Equations*, Pitman Research Notes in Math. Ser. II, John Wiley and Sons, New York, 1994-1995.

99. S. Zaidman, *An existence Theorem for Bounded Vector-Valued Functions*, Annali della Scuola Normale Superiore di Pisa, Classe di Scienze, 24, Fasc. 1 (1970), pp. 85–89.

100. S. Zaidmann, *Topics in Abstract Differential Equations* Nonllinear Analysis, Theory, Methods and Appl. **223** (1994), pp. 849–870.

101. M. Zaki, *Almost Automorphic Solutions of certain Abstract Differential Equations*, Annali. di Mat. Pura ed Appl., series 4, **101** (1974), pp. 91-114.

Index